Oxford Lecture Series in
Mathematics and its Applications 11

Series editors
John Ball   Dominic Welsh

OXFORD LECTURE SERIES IN MATHEMATICS AND ITS APPLICATIONS

1. J. C. Baez (ed.): *Knots and quantum gravity*
2. I. Fonseca and W. Gangbo: *Degree theory in analysis and applications*
3. P.-L. Lions: *Mathematical topics in fluid mechanics, Vol. 1: Incompressible models*
4. J. E. Beasley (ed.): *Advances in linear and integer programming*
5. L. W. Beineke and R. J. Wilson (eds): *Graph connections: Relationships between graph theory and other areas of mathematics*
6. I. Anderson: *Combinatorial designs and tournaments*
7. G. David and S. W. Semmes: *Fractured fractals and broken dreams*
8. Oliver Pretzel: *Codes and algebraic curves*
9. M. Karpinski and W. Rytter: *Fast parallel algorithms for graph matching problems*
10. P.-L. Lions: *Mathematical topics in fluid mechanics, Vol. 2: Compressible models*
11. W. T. Tutte: *Graph theory as I have known it*

Graph Theory
As I Have Known It

W. T. Tutte

Emeritus Professor
Department of Combinatorics and Optimization
University of Waterloo, Ontario

CLARENDON PRESS · OXFORD
1998

Oxford University Press, Great Clarendon Street, Oxford OX2 6DP
Oxford New York
Athens Auckland Bangkok Bogota Bombay
Buenos Aires Calcutta Cape Town Dar es Salaam
Delhi Florence Hong Kong Istanbul Karachi
Kuala Lumpur Madras Madrid Melbourne
Mexico City Nairobi Paris Singapore
Taipei Tokyo Toronto Warsaw
and associated companies in
Berlin Ibadan

Oxford is a registered trade mark of Oxford University Press

Published in the United States
by Oxford University Press, Inc., New York

© W. T. Tutte, 1998

All rights reserved. No part of this publication may be
reproduced, stored in a retrieval system, or transmitted, in any
form or by any means, without the prior permission in writing of Oxford
University Press. Within the UK, exceptions are allowed in respect of any
fair dealing for the purpose of research or private study, or criticism or
review, as permitted under the Copyright, Designs and Patents Act, 1988, or
in the case of reprographic reproduction in accordance with the terms of
licences issued by the Copyright Licensing Agency. Enquiries concerning
reproduction outside those terms and in other countries should be sent to
the Rights Department, Oxford University Press, at the address above.

This book is sold subject to the condition that it shall not,
by way of trade or otherwise, be lent, re-sold, hired out, or otherwise
circulated without the publisher's prior consent in any form of binding
or cover other than that in which it is published and without a similar
condition including this condition being imposed
on the subsequent purchaser.

A catalogue record for this book is available from the British Library

Library of Congress Cataloging in Publication Data
(Data available)
ISBN 0 19 850251 6

Typeset using LaTeX

Printed in Great Britain by
Bookcraft (Bath) Ltd
Midsomer Norton, Avon

FOREWORD

In 1984, just before his retirement from his regular duties at the University of Waterloo, Professor Tutte gave a fascinating series of lectures with the same title as this book. As one of the fortunate few who were able to attend those lectures, I have felt that a book based on them would be an invaluable source of information to all those who are interested in Graph Theory. I am very happy that such a book is at last appearing in print.

Professor Tutte is one of the principal pioneers of Graph Theory, an area in which he has been active for over sixty years. Thus what we have here is an enthusiastic first-hand account of many of the important developments in the subject. In addition, because of the autobiographical style of this work, we also have the privilige of glimpsing the creative process that led Professor Tutte to some of his famous discoveries.

The book is as entertaining as it is informative. The chapters bear subtitles taken from fairy tales, and have the quality of a magical narrative. Very few mathematicians have Professor Tutte's compelling story-telling ability.

I would like to thank Professor Dominic Welsh for recognizing the importance of having this book published, and Ms Elizabeth Johnston and Dr Julia Tompson of the Oxford University Press for their efforts in expediting its publication.

Waterloo, Ontario U.S.R. Murty
October 1997

PREFACE

This book is based on a course of lectures delivered at Waterloo, Ontario in 1984. Forgotten by the author they were brought to light again by the persuasive insistence of Professor U.S.R. Murty, and under his leadership prepared for publication.

The book is not presented as a comprehensive treatise on Graph Theory, being restricted to those parts of the theory that have been of special concern to the author. It is historical in the sense that it tells of how the author was led to his theorems and of the proofs that he used. Later and perhaps better proofs may only be referenced.

Besides Professor Murty the author wishes to thank Marg Feeney and Kim Gingerich for much demanding secretarial work. Making a readable text out of the preserved lecture notes was no easy task.

Waterloo, Ontario W.T.T
August 1997

CONTENTS

1	Squaring the square	1
2	Knights errant	12
3	Graphs within graphs	24
4	Unsymmetrical electricity	34
5	Algebra in Graph Theory	46
6	Symmetry in Graphs	64
7	Graphs on spheres	81
8	The Cats of Cheshire	94
9	Reconstruction	106
10	Planar enumeration	114
11	The chromatic eigenvalues	129
12	In conclusion	140
Bibliography		147
Index		153

1

SQUARING THE SQUARE

This book is about "Graph Theory as I have known it". This means that I propose to make my task easier by discussing mainly my own work and the parts of Graph Theory that have most interested me.

I begin therefore by trying to remember my own introduction to the subject. I put this introduction back in my high school days, when I found Rouse Ball's *Mathematical recreations and essays* in the school library. So I went on to university, in 1935, with some conception of a proof of the Five Colour Theorem, and with some knowledge of Petersen's Theorem, although not of its proof. But this did not make me a graph theorist, any more than my exposure to elementary calculus made me a mathematician. Indeed at Cambridge I read for the Natural Sciences Tripos, specializing in Chemistry. But there are graphs in Chemistry: first approximations to the structures of organic molecules.

I did have an interest in mathematical problems, strong enough to make me join the Trinity Mathematical Society. I often talked about such problems with three other members of the Society, students of Mathematics. They were Leonard Brooks, Cedric Smith and Arthur Stone. Each was destined to make his mark on Graph Theory. Who has not heard of Brooks' Theorem on the colouring of graphs, or of Smith's Theorem on Hamiltonian Circuits? Do I see a hand up? Never mind, you shall learn about these things in the course of these lectures. Stone is perhaps best known among combinatorialists for his theory of flexagons.

Stone found a problem in a book called *The Canterbury puzzles*, by H.E. Dudeney. One of these puzzles concerned a lady's "casket" with a square lid. There was a pattern on this lid whereby it was subdivided into a rectangle and a number of squares, all of different sizes. As I recall the puzzle no further information was given, except for the dimensions of the rectangle and its chemical composition. It was made of gold and it measured ten inches by one quarter of an inch. But one was asked to infer how many squares there were (remembering that they had to be all of different sizes), what those sizes were, and how the squares and the rectangle must fit together.

Stone was interested not so much in the solution that Dudeney gave as in the accompanying statement that the solution was unique. Unfortunately Dudeney gave no proof of uniqueness, nor any reference to such a proof.

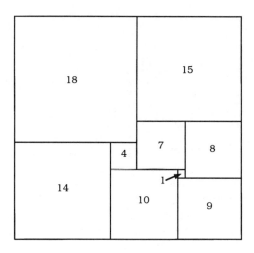

Figure 1.1 A squared rectangle.

But Stone thought that he could draw some remarkable consequences from the assertion of uniqueness. In particular no square could be dissected into smaller squares, all of different sizes. For if such a dissection were applied to the smallest of the lady's squares a second solution of the problem would be obtained.

Stone now found himelf with a deep mathematical problem. Is it possible to dissect a square into smaller squares, all unequal? On the basis of Dudeney's statement the answer was "No", and Stone set out to prove this. He had in mind, I think, that by success in this task he would prove himself as a mathematician. However he was not left alone to the work, for the other three members of the association had become interested too.

We found no proof in the mathematical literature, but we did find a few items of relevant information. Stone's problem had already arisen. The statement that the dissection was impossible was called Lusin's Conjecture, and it seemed to be only a conjecture. One book of recreations gave a dissection of a square into smaller squares with only one pair of equal ones. And Rouse Ball's book, already familiar to me, gave a dissection of a 32×33 rectangle into 9 unequal squares, as shown in Figure 1.1 That dissection was due to a Polish mathematician named Z. Moroń.

How were we to go on from there? All we could think of at this stage was to search for further "perfect rectangles", that is, rectangles dissected into unequal squares. After some practice we found that the construction of such rectangles was not difficult. Our method was to draw a rectangle dissected into smaller rectangles, and to pretend that these rectangles were badly drawn squares. On this assumption the relative sizes of the squares could

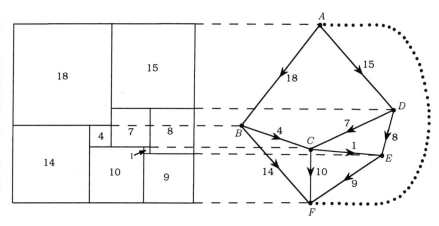

Figure 1.2 The electrical network of a squared rectangle.

be found by solving algebraic equations describing how the squares had to fit together. Sometimes two squares came out equal. That was bad luck, but it did not often happen unless there was some overlooked symmetry. Sometimes a square had to be given a negative side. But we found how to adjust the sign in those cases by a simple change of pattern. Very often the method worked. We amassed quite a respectable catalogue of perfect rectangles.

The first of the new perfect rectangles was found by Stone. It was of the 11th order, that is, it was dissected into 11 unequal squares. As with all subsequent perfect rectangles the side-lengths of the squares were commensurable. We could therefore take them to be all integers, and we could divide by their common factor to get them in lowest terms. The sides of Stone's rectangle were then 176 and 177. Stone remarked, sardonically, that it was nearly square, and there was not much further to go. Our other examples were mostly of the 9th, 10th and 11th orders. Higher orders required more difficult calculations. Alas, no perfect square made its way into our catalogue.

So far there was no suggestion of graph theory. But graphs began to appear when we tried to represent our rectangles by simplified diagrams. Eventually we found that a squared rectangle was equivalent to an electrical network of unit resistances. The equivalence is made plain on the cover of the Undergraduate Handbook of the Combinatorics and Optimization Department of the University of Waterloo. A similar diagram appears in Figure 1.2.

The horizontal lines in the squared rectangle correspond to the terminals of the network, and the squares correspond to the wires joining them. The current in a wire is measured by the side-length of the corresponding square, and its direction is downward in the rectangle. The top edge of the squared rectangle corresponds to the positive pole, the terminal at which current enters the network. The bottom edge likewise corresponds to the negative pole, the terminal from which the current leaves. The magnitude of the current entering at the positive pole and leaving at the negative can be equated to the length of a horizontal side of the rectangle. The voltage drop from pole to pole then measures the vertical side.

These simple observations brought in a flood of graph theory. A textbook of that theory may start by defining a graph as a set of objects called vertices, some pairs of which are joined by elements called edges. Accordingly our electrical network is a graph, with the terminals as vertices and the wires as edges. It is a "connected graph", that is, one can get from any vertex to any other by following edges. It is also a "planar graph", that is, it can be drawn in the plane without edge-crossings.

The construction of Figure 1.2 can be repeated with vertical segments instead of horizontal ones, as in Figure 1.3. It then gives a second graph or electrical network representing the same squared rectangle. In most cases the relation between the two graphs is clarified by a combinatorial and topological relation called "duality". In explaining this clarification it is best to adjoin to each network an extra wire joining the two poles. The extra wires appear as broken lines in Figures 1.2 and 1.3. On adjunction of its new wire each network becomes a "completed net" or "c-net" of the squared rectangle.

Each c-net dissects the plane into disjoint connected regions called the "faces". To say the two c-nets are dual means that one can draw them together in the plane so that each vertex of either c-net lies inside a face of the other, so that each face of either c-net contains exactly one vertex of the other, and so that each edge of either c-net meets the other c-net in one point only, an internal point of its corresponding edge.

The c-nets of Figures 1.2 and 1.3 are drawn together in this way in Figure 1.4. The arrows on the new or "polar" edges go from the positive pole to the negative.

I stated that the above account applies "in most cases". The implied exceptions are those perfect rectangles that have a "cross", a point at which four squares meet, corner to corner. At such a point one has a choice in specifying the horizontal segments. One may consider that there is just one horizontal segment at the cross, having the cross as one of its internal points. Alternatively one may recognize two horizontal segments at the cross, meeting end to end. An acceptable electrical network can be constructed on either assumption. To get dual electrical networks one must have at each cross either a single horizontal segment and two vertical ones or a single vertical and two horizontal. Let it be said however that it is quite difficult to construct a perfect rectangle with a cross.

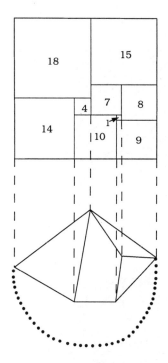

Figure 1.3 A second electrical network.

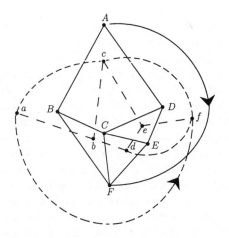

Figure 1.4 Dual c-nets.

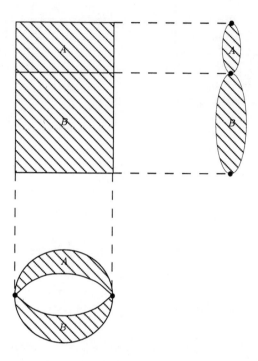

Figure 1.5 Sketch of a trivially compound rectangle and its networks.

We four students found ourselves struggling with a rather advanced part of graph theory, that of planarity and planar duality. Soon we had to cope with the theory of 3-connection as well.

Given two perfect rectangles we could draw them with their horizontal sides of equal length and then stack one upon the other to make, with reasonable luck, a third. But we looked upon this construction as trivial, and we did not list such combinations in our catalogue. I sketch an example in Figure 1.5, and sketch with it its two electrical networks.

The figure indicates that from the horizontal segments, and barring crosses, we get a network that is in two parts, having only one vertex in common. In the other network there are two parts having only two vertices in common. In a standard terminology a graph is called "separable" if it can be dissected into two parts having only one vertex in common, provided that neither part consists solely of that one vertex. A non-separable connected graph is called "2-connected".

A 2-connected graph made up of two parts with just two vertices in common, and such that each part has at least two edges not belonging to the other, is often called "2-separable". A 2-connected graph that is not 2-separable is "3-connected".

Of the two electrical networks of Figure 1.5 one is separable and the other 2-separable. It can be shown that the corresponding c-nets are both 2-connected.

The trivial squared rectangle of Figure 1.5 is an example of a "compound rectangle", one that contains a smaller squared rectangle. Squared rectangles that are not compound are "simple". The four researchers, somewhat arbitrarily perhaps, agreed to list only simple perfect rectangles in their catalogue. They soon satisfied themselves that the c-net of a simple perfect rectangle is necessarily 3-connected.

Planar 3-connected graphs have a fascinating theory. A theorem of Hassler Whitney says that each of them can be drawn in the plane, or rather on the sphere, in essentially only one way. A theorem of Steinitz identifies these graphs with those determined by the edges and vertices of the convex polyhedra.

We four students came upon this much graph theory even before we considered the electrical properties of our networks. But a discussion of those properties required even more. We now understood that the basic theory of squared rectangles was that of Kirchhoff's Laws of electrical networks. That was something we could look up in textbooks. We referred to Jeans' *Electricity and magnetism*. We learned that Kirchhoff's Laws for a network N were associated with a special matrix $K(N)$ called, by us at least, the Kirchhoff matrix of N. It was convenient to enumerate the vertices of N as $v_1, v_2, \ldots, v_n$. Then the jth diagonal element c_{jj} of $K(N)$ could be defined as the sum of the conductances of the wires joining v_j to other vertices. For distinct suffixes i and j the element c_{ij} in the ith row and the jth column had to be equal to minus the sum of the conductances of the wires joining v_i and v_j. In the cases of interest to us the conductance of each wire was unity, and each non-diagonal c_{ij} was either 0 or -1.

By its definition the matrix $K(N)$ was symmetrical, and the elements in each row and column summed to zero. Thus

$$\det K(N) = 0. \tag{1.1}$$

We learned that the determinant of the matrix $K_i(N)$ obtained from $K(N)$ by striking out the ith row and column was independent of i. We decided to call its value the "complexity" $C(N)$ of N. We showed, with some initial difficulty, that it was positive for all connected N, although zero for all disconnected networks. We found it convenient to take $C(N)$ as a measure of the total current entering at the positive pole v_p and leaving at the negative pole v_q. Then the potential drop between the two poles was, by the theory of the textbook, the determinant of the matrix obtained from $K(N)$ by striking out the rows and columns of the two poles. Other determinants derived from $K(N)$ gave the potential drops between specified pairs of vertices, and thus the currents in individual wires.

From this information we inferred that if the horizontal side of a squared rectangle was made equal to the complexity of the corresponding electrical network, then both sides of the rectangle and the sides of all the component squares would be integers. We called these integers the "full sides" and "full elements" respectively of the rectangle. Their highest common factor ρ we called the "reduction". Dividing by the reduction we got the "reduced" sides and elements. Most of the perfect rectangles in our catalogue had reduction 1. But other reductions did occur.

Some coincidences in our catalogue of simple perfect rectangles were now explained. Two rectangles whose uncompleted electrical networks were identical except for the choice of poles necessarily had the same full horizontal side. Two rectangles whose uncompleted electrical networks became identical when the two poles were identified would have the same full vertical side. To these observations we were able to add that if two rectangles had the same completed network with a different choice of polar edge then they had the same full semiperimeter, the same sum of vertical and horizontal sides. This sum would be the complexity of the c-net.

We wondered if it was possible to construct two perfect rectangles that could be shown by these rules to have the same full horizontal side and the same full vertical side. The story of how we eventually found such a pair is long and involved, but parts of it have been found amusing [58]. (See Chapter 12). I cut it short by asking you to contemplate the two diagrams of Figure 1.6.

These can be interpreted as electrical networks of squared rectangles, A being the positive pole and C the negative in each case. Since they are identical, except for the choice of poles, we can assert that the two rectangles have the same full horizontal side. Moreover the completed networks are identical. As with the uncompleted networks, one can be transformed into the other by a rotation through 120 degrees. Hence the full semiperimeters of the two rectangles are equal. Calculations show that the two congruently shaped rectangles are indeed different, and each is perfect. Moreover no full element of one rectangle, with one exception, appears as a full element of the other. The exception corresponds to the wires marked x. These are found to carry equal full currents, and so to correspond to equal squares. Conveniently these turn out to be corner squares in the two rectangles.

The two perfect rectangles corresponding to Figure 1.6 can be put together to make a perfect square, as indicated in Figure 1.7. They are made to overlap in their common corner square x and two other squares, shaded in the diagram, are added.

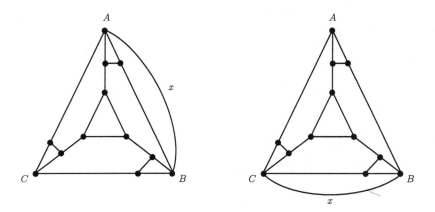

Figure 1.6 Use of rotational symmetry.

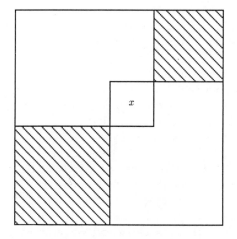

Figure 1.7 Overlapping rectangles.

In the above construction we see one example of a general method. In general we need a planar graph with 3-fold rotational symmetry, but with no line of symmetry. Given such a graph drawn in the plane we distinguish three symmetrically equivalent vertices A, B and C on the outer boundary. I refer to such a figure as a "rotor", and to A, B and C as its "vertices of attachment". We get two electrical networks by joining A to B in one case, and B to C in the other, by a new edge x and then taking A and C as the poles. With luck there will be no equalities between the full elements of the two rectangles except for the pair of squares corresponding to the edge

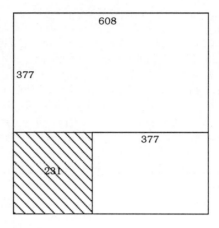

Figure 1.8 Sketch of a compound perfect square.

x, and in that case the rectangles can be put together to make a perfect square. Note that any perfect square constructed in this way must have at least two crosses.

We four researchers made further generalizations. We generalized the notion of a rotor from 3-fold to n-fold rotational symmetry. Then by various tricks we were able to construct simple perfect squares, and even simple perfect squares without crosses ([11], [58]).

While this was going on we extended our catalogue, finding all the 3-connected planar graphs, with no double joins, of up to 13 edges, and many more of 14 edges. For most of these we calculated the squared rectangles for which they were c-nets. Soon the catalogue listed more than 200 perfect rectangles. We noticed that it was possible to fit some pairs of these rectangles together, usually with an extra square or two, to make "empirical" perfect squares. Thus our paper of 1940 records a square of the 26th order obtained by fitting together two perfect rectangles (377 × 608 and 231 × 377) and a square of side 231, as indicated in Figure 1.8.

The first perfect square to be published was not ours, but it was produced by the empirical method of putting together perfect rectangles from a catalogue, in this case more than two. It was constructed by R.P. Sprague, who published it in *Mathematische Zeitschrift* in 1939. It was of the 55th order ([55], [56]).

It is distressing in a way that the empirical method should get better results, that is perfect squares of smaller order, than the symmetrical or "theoretical" method. In 1940 our best "theoretical" squares were of the 39th order. Later T.H. Willcocks found an empirical perfect square of the 24th order and a theoretical one of the 37th ([119], [120]).

The cataloguing of squared rectangles is now computerized, and they have been listed up to the 21st order. And one in the 21st order is a simple perfect square [22]. This is now known to be the unique perfect square of lowest order ([21], [23], [41]). Oldthinkers fondly hope that a theoretical explanation of this and other low-order perfect squares may yet be achieved.

The immediate purpose of this discussion is to show how the study and eventual disproof of Lusin's Conjecture was for me a most effective introductory course in Graph Theory. One more remark should be made in this connection, concerning the spanning trees of a graph. A tree is usually defined as a connected graph that has no circuit. Equivalently we can say that a tree is a connected graph that is disconnected by the removal of any edge. A "spanning tree" of a graph G is a tree that is a subgraph of G including all the vertices of G. By expanding a determinant it can be shown that the number of spanning trees of G is the complexity $C(G)$ of G. For a general electrical network N the complexity $C(N)$ proves to be the sum over all the spanning trees of N of the products of the conductances of their edges. This result is known as the Matrix–Tree Theorem, and it goes back to Kirchhoff. An associated result says that dual planar graphs, of unit resistances, have equal complexities, that is equal tree-numbers.

Up to 1940 Graph Theory was not much studied. It is thus not really surprising that the four researchers now found themselves looked upon as leading experts in that subject. In my own case I was moved to turn from Chemistry to Mathematics and to the writing of papers on Graph Theory. Some of these papers will be discussed in the following chapters.

2
KNIGHTS ERRANT

Let me remind you of the ancient game of chess. There is a board of 64 squares, eight by eight. There are pieces with names like "King", "Bishop" and "Pawn", and each piece has its own set of permissible moves.

The game generates some well-known combinatorial problems. For example, how many Queens can be put on the board so that no one is on a square controlled by another? The game itself is a big combinatorial problem: does either Black or White have a certain win from the initial position, given perfect play by both sides? No answer to that one is expected in this generation. The chess problem that concerns me in this lecture is that of the Knight's Tour.

The Knight's move is two squares vertically followed by one horizontally, or two horizontally and one vertically. The problem is to plan a tour of the board for the Knight whereby he returns to his original position after visiting each other square of the board once and once only.

Some solutions of the problem are given in Rouse Ball's book. I now use the thirteenth edition of that work, co-authored by H.S.M. Coxeter [2]. Therein I read that the earliest solutions known to the author, or authors, are those given by De Montmort and De Moivre in the early eighteenth century. But the first serious attempt to deal with the problem by mathematical analysis was made by Euler in 1759. Rouse Ball's book gives a sketch of Euler's method and presents one of his solutions.

The problem of the Knight's Tour is one in Graph Theory. The board reduces to a graph, of which the 64 squares are the vertices. Two of these are joined by an edge whenever they are a Knight's move apart. A Knight's re-entrant tour can be interpreted as a subgraph of this graph G, a subgraph obtained by deleting all the edges of G not traversed in the tour. This subgraph is spanning; that is, it contains all the vertices of G. It is also a circuit; that is, a connected graph of valency 2, a connected graph in which each vertex is incident with exactly two edges. Nowadays we generalize the Knight's problem in the following question: when and how can we find a spanning circuit in a given graph? It seems that Euler did not attempt that full generalization, but he did solve the Knight's problem for some chessboards of non-standard shape.

A century later there came upon the scene another Knight, Sir William Rowan Hamilton, discoverer of quaternions and an enthusiastic student of the regular dodecahedron. He observed that the graph defined by the edges and vertices of that solid had spanning circuits, and he tried to popularize them with his "Icosian Game". For that story I refer you to the historical work of N.L. Biggs, E.K. Lloyd and R.J. Wilson, entitled *Graph Theory 1736–1936* [7].

In that book you can read also of the work of T.P. Kirkman who, at about the same time, studied spanning circuits on general convex polyhedra. There are questions of priority and propriety here. Biggs, Lloyd and Wilson remark that Hamilton was concerned with one special case, whereas Kirkman discussed general polyhedra. Moreover they point out that Kirkman was the first to publish. Nevertheless the spanning circuits of a graph are now known as its Hamiltonian circuits, not as its Kirkman circuits. If we wished to be entirely fair in our terminology we would no doubt hark back to the problem of the Knight's tour and consider calling them Euler paths (of the second kind?). And what about De Montmort and De Moivre?

In the eighties of the nineteenth century we find the physicist P.G. Tait much concerned about the relationship between Hamiltonian circuits and the Four Colour Problem. This was in the interval between Kempe's "proof" of 1879 and Heawood's refutation of 1890. Tait acknowledged Kempe's proof but sought for new ones. The relationship is simply that if a planar map has a Hamiltonian circuit in its graph then its faces can be 4-coloured. Those inside the circuit can be coloured alternately red and blue, and those outside alternately green and yellow.

Tait confined his attention to the cubic or trivalent planar maps, those in which just three edges meet at each vertex. This was because the Four Colour Problem had already been reduced to that case. Indeed the problem had been further reduced to the 3-connected case. Anticipating Steinitz' Theorem, Tait took the 3-connected planar cubic graphs to be equivalent to graphs defined by the edges and vertices of convex polyhedra. He referred to them and their corresponding maps as "true polyhedra". A cubic planar graph that was only 2-connected would presumably be a false polyhedron. (A "map" is distinguished from a graph by having faces as well as edges and vertices.)

Tait conjectured that every true (cubic) polyhedron has a Hamiltonian circuit. Biggs, Lloyd and Wilson tell us that Tait made his conjecture in 1880 and that he consulted Kirkman about it. Kirkman concluded that Tait's Conjecture "mocks alike at doubt and proof".

Let us move along to the 1930s and to the four "Squarers of the Square" at Trinity College, Cambridge. They did not confine their mathematical recreation to that one problem. Smith heard of Tait's Conjecture and proposed it as something else that merited our attention. Our studies led eventually to a disproof of the Conjecture, published in 1946 [60]. I propose now to tell you of that disproof. If I may adapt a claim of the White

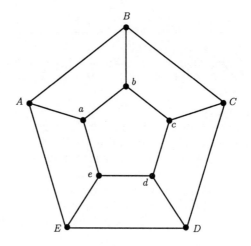

Figure 2.1 The pentagonal prism.

Knight, "It's long, but it's very, very beautiful."

Conformally with our procedure in the case of squared rectangles we began by making a catalogue of 3-connected planar cubic maps and recording their Hamiltonian circuits. We got little general information from this, beyond counter-examples to some of our wilder conjectures.

One of these counter-examples proved to be of recurring interest. We had conjectured that in any 3-connected planar cubic map we could find a Hamiltonian circuit through two arbitrarily chosen edges. But this conjecture failed for the pentagonal prism, shown in flattened form in Figure 2.1.

It is easy to verify that this map has no Hamiltonian circuit through the two edges Aa and Cc.

We tried a weaker conjecture, that there is a Hamiltonian circuit through any two adjacent edges. I was then able to score one of our first successes in this theory, by showing that this proposition was a consequence of Tait's Conjecture.

The proof begins with the assumption that Tait's Conjecture is true and that the new one is not. Then there is a 3-connected cubic planar map M with a vertex v and its three incident edges A, B and C such that there is no Hamiltonian circuit of M through both B and C. This means that all the Hamiltonian circuits of M pass through the edge A.

Let us detach the edges A, B and C from v, deleting that vertex and giving A, B and C new ends v_A, v_B and v_C respectively in its place. The map M is then changed into the structure M' shown in Figure 2.2. Let us call v_A, v_B and v_C the "outer" vertices of M', and the remaining vertices

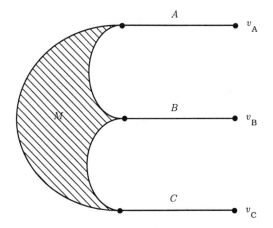

Figure 2.2 A stage in an extension of Tait's Conjecture.

of M the "inner" ones.

The statement that every Hamiltonian circuit of M goes through A becomes the following statement about M': If a path in M' goes from one outer vertex to another, and passes through all the inner vertices, then it traverses A, beginning or ending at v_A.

The trick is to take three copies of M', say M'_1, M'_2 and M'_3, and put them together as in Figure 2.3. Edges A, B and C belonging to the copy M'_i are distinguished by the suffix i.

The figure so constructed can be seen to be a 3-connected cubic planar map. By our assumption of Tait's Conjecture the map has a Hamiltonian circuit H. The part of H in M'_i joins two outer vertices and passes through all the inner ones. So H passes through A_i, by our observation about M'. Since no circuit of the map can pass through all three of the adjacent edges A_1, A_2 and A_3 we now have a contradiction. Accordingly Tait's Conjecture implies the existence of a Hamiltonian circuit through any two adjacent edges.

In later years I made two successive improvements on this result. Both were simple and the second led immediately to a disproof of Tait's Conjecture. Looking back I am surprised at how long a time it took for me to reach this conclusion. We had the result just described by 1940 but I got to the final step only in 1942 or 1943, and did not write my results for publication until 1945. Of course, up to the last months of that year I had more pressing business.

The first strengthened form asserts that if Tait's Conjecture is true then there exists a Hamiltonian circuit through any two edges bordering the same face. To prove it, consider any face F of a map M incident with

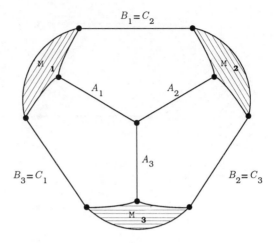

Figure 2.3 Second stage in the extension.

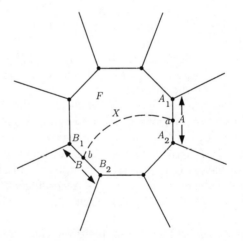

Figure 2.4 Second extension of Tait's Conjecture.

two specified edges A and B, as in Figure 2.4.

We subdivide A into two edges A_1 and A_2 meeting at an internal point a of A, as shown in the figure. We then make an analogous subdivision of B, at an internal point b. We next join a and b by a new edge X crossing F. We find that this operation converts M into a new 3-connected planar cubic map M_0. By the preceding result M_0 has a Hamiltonian circuit H_0

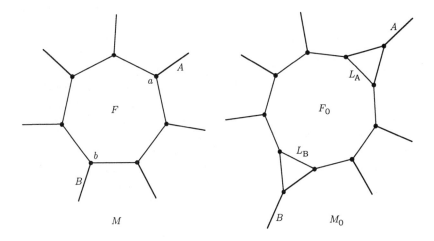

Figure 2.5 Third extension.

passing through A_1 and A_2. It follows that H_0 does not traverse X, and therefore that it passes through both B_1 and B_2. Clearly H_0 derives from a Hamiltonian circuit H of M that passes through both A and B. Our first "strengthened form" is now proved.

Let us now define the "radiants" of a face F as the edges having an end incident with F but not themselves incident with F, such as A and B in Figure 2.5. My second strengthened form runs as follows. Let A and B be radiants of a face F. Then if Tait's Conjecture is true there is a Hamiltonian circuit of the map that passes through both A and B.

The trick in this case is to transform M into a new map M_0 by expanding two vertices into triangular faces, as in Figure 2.5

The vertices so treated are the ends a and b of A and B respectively that are incident with F. Having satisfied ourselves that M_0 is a planar 3-connected cubic map we can infer, using the first strengthened form, that it has a Hamiltonian circuit H_0 passing through the sides L_A and L_B of the two new triangles opposite A and B respectively.

It is now easy to see that H_0 must traverse both A and B. Moreover when the construction is reversed by contracting the two triangles back to vertices, H_0 contracts into a Hamiltonian circuit H of M that passes through both A and B. The second strengthened form is proved.

In this second strengthened form we have proved altogether too much for Tait's Conjecture. We have already remarked that the graph of the pentagonal prism has two edges that cannot lie in the same Hamiltonian circuit. These are the edges Aa and Cc of Figure 2.1, and they are radiants of each pentagonal face. The assumption of Tait's Conjecture has led us to a contradiction, and therefore that conjecture is false. It is not difficult to construct a counter-example to Tait's Conjecture by starting with the pentagonal prism and reversing the constructions of the above reasoning. I presented this counter-example in my paper of 1946.

Smith's work at Trinity on Hamiltonian circuits was also fruitful. It led him to an algebraic theory of "3-nets", not published until 1971 [54]. It led him also to Smith's Theorem, the assertion that in any cubic graph, planar or non-planar, 3-connected or not 3-connected, the number of Hamiltonian circuits through any specified edge is even. Those who apply this theorem should bear in mind that zero is an even number. Smith's Theorem has the intriguing corollary that if a cubic graph has one Hamiltonian circuit it has at least three. Later I found a proof shorter than Smith's, and I included this proof in the paper of 1946.

A cubic graph is called "cyclically n-connected" if it cannot be split into two parts, each containing a circuit, by deleting fewer than n edges. My counter-example was cyclically only 3-connected; it seemed right to search further for cyclically 4-connected and 5-connected counter-examples to Tait's Conjecture. (There are no cyclically 6-connected cubic planar maps if we exclude trivialities such as the tetrahedron). In 1960 I was able to publish a cyclically 4-connected counter-example [62]. A cyclically 5-connected one, due to H. Walther, appeared in 1965 [109].

A theorem of the Latvian mathematician E. Grinberg, published in 1968 [31], made the discovery of non-Hamiltonian planar maps very much easier, and produced very much simpler cyclically 5-connected counter-examples. It is a theorem with a simple proof, so simple that one marvels that it was not discovered long ago, say by Tait or Kirkman. Grinberg's Theorem is usually applied to cubic maps, but it is really a theorem about planar maps in general.

Let M be a planar map (drawn on a sphere), and suppose it to have a Hamiltonian circuit H. As a circuit, H separates the rest of the sphere into two simply connected regions that we may as well call the inside and outside of H. Let us write h for the number of edges of H, d' for the number of other edges crossing the inside of H, and d'' for the number crossing the outside. Let f'_i and f''_i denote the number of i-gons inside and outside H respectively.

It is easily seen that the number of faces inside H is $d' + 1$, and the number outside is similarly $d'' + 1$. Thus

$$\sum_i f'_i = d' + 1, \qquad (2.1)$$

$$\sum_i f''_i = d'' + 1. \qquad (2.2)$$

We see also that
$$\sum_i i f'_i = h + 2d', \tag{2.3}$$

$$\sum_i i f''_i = h + 2d''. \tag{2.4}$$

Hence
$$\sum_i (i-2) f'_i = h - 2, \tag{2.5}$$

$$\sum_i (i-2) f''_i = h - 2, \tag{2.6}$$

and therefore
$$\sum_i (i-2)(f'_i - f''_i) = 0. \tag{2.7}$$

This is Grinberg's Theorem.

Consider a planar map M in which the number of i-gons is f_i. It may be possible to show that equation (2.7) is false whenever each f_i is partitioned into two non-negative integers f'_i and f''_i. If so it will follow that M has no Hamiltonian circuit.

One of Grinberg's non-Hamiltonian maps is shown in Figure 2.6. There the number written inside each face is the number of its edges. There are 21 pentagonal faces, 3 octagons and a single nonagon. Equation (2.7), reduced mod 3, now takes the simple form

$$f'_9 - f''_9 \equiv 0 \pmod{3}. \tag{2.8}$$

But this congruence is false, since one of the numbers f'_9 and f''_9 has to be 1 and the other 0. This argument proves that the map has no Hamiltonian circuit.

It can be verified that the map of Figure 2.6 is cyclically 5-connected. Another cyclically 5-connected counter-example to Tait's Conjecture, also due to Grinberg, is shown in Figure 2.7. It has 24 faces, as against 25 for the map of Figure 2.6. Both maps are very much smaller than the cyclically 4-connected counter-example of 1960 and the cyclically 5-connected one of 1965. The proof of non-Hamiltonicity for the map of Figure 2.7 is a little more complicated than for that of Figure 2.6.

The new map has 18 pentagons, 3 hexagons and 3 octagons. From equation (2.7), reduced mod 3, we have

$$f'_6 - f''_6 \equiv 0 \pmod{3}. \tag{2.9}$$

Since the two numbers f'_6 and f''_6 sum to 3 we may assume without loss of generality that $f'_6 = 3$ and $f''_6 = 0$. Accordingly the three central hexagons

20　　　　　　　　　　KNIGHTS ERRANT

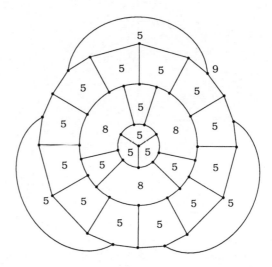

Figure 2.6 A non-Hamiltonian cubic graph.

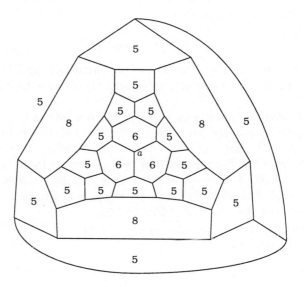

Figure 2.7 Another non-Hamiltonian cubic graph.

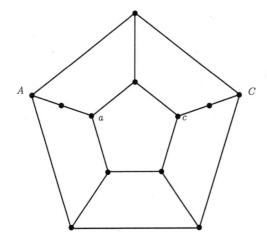

Figure 2.8 An application of Grinberg's Theorem.

are all inside the assumed Hamiltonian circuit. But this is impossible since H is required to pass through their common incident vertex a. So in fact the map has no Hamiltonian circuit.

It is interesting to consider the map of Figure 2.1 in the light of Grinberg's Theorem. The statement that it has no Hamiltonian circuit through both edges Aa and Cc is equivalent to the assertion that the non-cubic map of Figure 2.8 has no Hamiltonian circuit. The faces of this map are 6 pentagons and one quadrilateral. From equation (2.7), reduced mod 3, we infer that

$$f'_4 - f''_4 \equiv 0 \pmod{3}. \tag{2.10}$$

But this is impossible since there is one and only one quadrilateral.

The theory of non-Hamiltonian cubic planar maps has been carried further by D. Younger and J. Zaks. But it is more in accord with the knightly chessboard tradition to prove existence, not non-existence, for Hamiltonian circuits. That is done for example in Dirac's Theorem about strict graphs. By a "strict graph" I mean a graph in which each edge has two distinct ends, and no two vertices are joined by more than one edge. In some contexts it is convenient to use more general graphs in which two or more edges may have the same pair of ends, and even graphs in which an edge called a "loop" joins some vertex to itself. Some writers call my "strict graphs" simply "graphs", and use terms like "multigraph" for the more general kinds. But in this book I shall call all these structures "graphs". I shall also use the term "valency", borrowed from Chemistry. The valency of a vertex in a graph is the number of incident edges, loops being counted twice.

The theorem of G.A. Dirac just mentioned asserts the following. Any strict graph with at least three vertices and in which the minimum valency is at least half the number of vertices has a Hamiltonian circuit [20]. This theorem, a stronger one due to L. Pósa, and some related results are discussed in the monograph of H. Walther and H.-J. Voss entitled *Über Kreise in Graphen* ([50], [111]).

A map on the sphere in which each face is triangular is called a "planar triangulation". I have found it convenient also to use the term "near-triangulation" for a map on the sphere in which every face, with at most one exception, is triangular. When I was at Cambridge I learned about a famous theorem of Hassler Whitney about Hamiltonian circuits in near-triangulations. It is usually stated in the following simplified form: any strict triangulation in which there is no separating triangle has a Hamiltonian circuit.

By a "separating triangle" of a planar map is meant a circuit of three edges that is not the boundary of a face; it has at least two faces of the map inside and at least two outside. In any planar map it is common to specify one of the faces as the "outer" face. Here I make the convention that in a near-triangulation the non-triangular face, if there is one, must be chosen as the outer face. The more complicated form of Whitney's Theorem, needed to make possible an inductive proof, runs as follows. Any near-triangulation without a separating triangle has a Hamiltonian circuit. Moreover this circuit can be chosen to pass through any two specified edges incident with the outer face [113].

In the early fifties, at the University of Toronto, I got the idea of trying to generalize Whitney's Theorem from near-triangulations to general planar maps. I recall that the crucial operation in this research was to work through Whitney's proof, changing each sentence in turn so as to make it of more general application. In particular I had to replace Hamiltonian circuits by a less restricted kind. I wish I could now remember more details of this research, for I have often been questioned about it. But those details seem irrecoverable.

Like Whitney's Theorem the generalization is usually stated in a simplified form: any 4-connected planar map has a Hamiltonian circuit. The term "4-connected" means that two non-adjacent vertices in the map are never separated by fewer than 4 others. This 4-connection is not the cyclic 4-connection that I spoke of earlier, but a dual concept. I sometimes distinguish it as "vertical" 4-connection. Let me now try to explain the strong form of the generalization.

Let C be any circuit of a graph G. Let its "residual graph" $R(C)$ be defined by the vertices of G not in C and the edges joining them. A "non-degenerate bridge" B of C is a graph defined by a component of $R(C)$, the edges joining it to C and their ends in C. These last-mentioned vertices

I call the "vertices of attachment" of B. The union of C and its non-degenerate bridges is not necessarily the whole of G; it leaves out any edges having both ends in C but not themselves belonging to C. It leaves out for example all loops having their "two coincident ends" in C. Let us say that any edge not in C but with both its ends in C constitutes, with its ends, a "degenerate bridge" of C. These ends can also be called the vertices of attachment of the degenerate bridge. For any bridge B of C in G, degenerate or non-degenerate, I write $w(B)$ for the number of vertices of attachment.

The general theorem applies to a planar map M defined by a graph G drawn in the plane. Only two restrictions are put on G; it must be connected and it must have at least one circuit. A circuit J has two residual domains R and S in the plane, and each bridge of J is contained, except for its vertices of attachment, in one or other of these. If either R or S contains no bridge with more than one vertex of attachment, it is a "terminal domain" of M, and J is a "terminal circuit". If M is 2-connected then the terminal domains are simply the faces. The general theorem says that any such map M has a circuit C such that no bridge of C in G has more than three vertices of attachment. Moreover there exists such a circuit C passing through any two specified edges of any specified terminal circuit. The inference that 4-connected maps have Hamiltonian circuits is not entirely trivial. It is based on the observation that a non-degenerate bridge B with $w(B) < 4$ destroys the possibility of 4-connection in most cases. The only non-trivial exceptional case is that in which C is a 3-circuit with a single bridge B, that bridge satisfying $w(B) = 3$.

The theorem was published in 1956 [61], and a second exposition followed in 1977 [63]. This theorem too is discussed in the Walther–Voss monograph. See also Horst Sachs, *Einführung in die Theorie der endlichen Graphen, Teil* II (Leipzig 1972).

3

GRAPHS WITHIN GRAPHS

We met with spanning trees of a graph in the first chapter and with spanning circuits in the second. We shall next be concerned with spanning subgraphs satisfying certain valency conditions.

Let k be any positive integer. Then a subgraph of a graph G is called a "k-factor" of G if it includes all the vertices of G and is a regular graph of valency k. By a "regular" graph we mean one in which all the vertices have the same valency, in this case k. A 1-factor of G is often called a "perfect matching". It can be regarded as a pairing of the vertices so that each pair is joined by an edge. Evidently G can have no 1-factor if the number of its vertices is odd. A Hamiltonian circuit of G is a connected 2-factor.

More generally we can introduce a function f from the set of vertices of G to the set of non-negative integers. We can then define an "f-factor" of G as a spanning subgraph of G in which the valency of v is $f(v)$ for each vertex v. There is an important branch of Graph Theory concerned with conditions for a given graph to have an f-factor, for a given function f.

1-factors became prominent with the work of P.G. Tait on the Four Colour Problem. He showed that for planar cubic maps the Four Colour Theorem was equivalent to the following proposition: the edges of the map can be coloured in three colours α, β and γ so that no two adjacent edges have the same colour. Such an edge-colouring of any cubic graph, planar or not, is called a Tait colouring.

In a Tait colouring of any cubic graph each of the three colours defines a 1-factor. The colouring is often described as a resolution of the graph into three 1-factors. The union of any two of the 1-factors is called a Tait cycle of the graph. It is a 2-factor of a special kind, one in which the number of edges of each component circuit is even. The Hamiltonian circuits of a cubic graph G are Tait cycles. Each of them is of even length since a cubic graph necessarily has an even number of vertices. From any one of them we can construct a corresponding Tait colouring by colouring its edges alternately α and β, and then assigning the colour γ to every remaining edge of G.

Biggs, Lloyd and Wilson [7] tell us that the first general theoretical discussion of the problem of factorizing graphs appeared in a paper of J. Petersen in 1891 [48]. Petersen showed that every regular graph of non-zero even valency can be resolved into 2-factors. More importantly he showed that every connected cubic graph with no more than two "leaves" has a

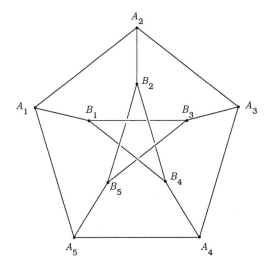

Figure 3.1 The Petersen graph.

1-factor. This is Petersen's Theorem. I usually restate it by saying that a connected cubic graph has a 1-factor if either it has no isthmus or all its isthmuses lie on a single arc. Here an "isthmus" is an edge whose deletion destroys the connection of the graph.

Petersen's Theorem made it clear that each of Tait's "true polyhedra" had a 1-factor. But that was a long way from showing that each had three mutually edge-disjoint ones, and thereby proving the Four Colour Theorem.

It is hard to find a cubic graph, without an isthmus, that has no Tait colouring. No cubic graph with an isthmus has a Tait colouring, but that fact is regarded as trivial. The first example was published by J. Petersen in 1898 [49]. It is the famous "Petersen graph", shown in Figure 3.1.

Note that the pentagon of A's and the pentagon of B's together make up a 2-factor of the graph. But this 2-factor cannot be a Tait cycle since its two component circuits are of odd length. The edges A_iB_i constitute a 1-factor, in accordance with Petersen's Theorem.

The Petersen graph is highly symmetric. It is in fact the graph obtained from that of the regular dodecahedron by identifying diametrically opposite points. It can be shown that all its arcs of length 3, directed arcs moreover, are equivalent under the symmetry. It can also be shown that the Petersen graph is the smallest cubic graph in which there is no circuit of fewer than five edges.

The use of these facts makes it fairly easy to show that the Petersen graph has no Tait colouring. If it had one it would have a Tait cycle. This would have 10 edges and each of its components would have to be a circuit of even length > 5. The Tait cycle would therefore be a Hamiltonian circuit. But we can show that the Petersen graph has no Hamiltonian circuit in the following way.

Suppose the graph had a Hamiltonian circuit H. By symmetry we may suppose it to contain the edges A_1A_2, A_2A_3 and A_3A_4. It must therefore exclude A_2B_2 and A_3B_3 and so contain all the edges of the arc $B_1B_3B_5B_2B_4$. Then, to avoid a short circuit it must exclude B_1B_4. Therefore it must include A_1B_1 and A_4B_4. But now H is seen to contain the circuit $A_4A_3A_2A_1B_1B_3B_5B_2B_4$. Since this does not pass through A_5 the definition of H as a Hamiltonian circuit is contradicted.

When I went up to the university I had heard of Petersen's Theorem and Petersen's Graph. When I began to study Graph Theory in connection with squared rectangles and Hamiltonian circuits I looked for more information about 1-factors. I found some in the library of the Cambridge Philosophical Society, in a paper by M.A. Sainte-Laguë entitled "Les réseaux (ou graphes)" [51]. There I found a proof of Petersen's Theorem based on Petersen's own, and I read it very carefully. There seemed to be a few gaps, due perhaps to over-condensation, but I was able to fill them to my satisfaction. I did not return to the subject until I had started my PhD research at Trinity (1945–1948). Then I got a new idea from a book on determinants.

I mentioned in the first chapter that potential differences in an electrical network could be expressed as determinants derived from the Kirchhoff matrix. Some very interesting results could be derived from quadratic identities depending on Jacobi's Theorem about the minors of an adjugate matrix. For example if the full horizontal side of a perfect square was written as pq^2, where p is square-free, then the reduction of that square would be at least pq, and the reduced side at most q. This necessity for a large reduction, I thought, explained why we had encountered no perfect squares within the range of our catalogue.

I browsed through some textbooks on determinants looking for other theorems that might be applicable to squared rectangles and squares. I came upon the explicit formula for a Pfaffian and I realized that it had something to do with the theory of 1-factors.

Let me remind you. Suppose we have a skew-symmetric matrix M of even order $2k$. If m_{ij} is the element in the ith row and jth column, then

$$m_{ij} = -m_{ji} \tag{3.1}$$

by the skew-symmetry. The determinant of M is known to be of the form P^2, where P is a polynomial in the m_{ij} called the Pfaffian of M. The explicit formula for P goes like this:

$$P = \sum \in m_{i_1j_1} m_{i_2j_2} \ldots m_{i_kj_k}. \tag{3.2}$$

The sum is over all possible arrangements of the integers from i to $2k$ in pairs (i_n, j_n). The order of the elements of each pair and the order in which the pairs are written are immaterial. The symbol $\in$ denotes the sign of the permutation
$$(i_1, j_1, i_2, j_2, \ldots, i_k, j_k)$$
of the sequence
$$(1, 2, 3, \ldots, 2k).$$

Clearly the terms of P are in 1-1 correspondence with the 1-factors of a complete graph of $2k$ vertices. The vertices can be enumerated as $v_1, v_2, \ldots, v_{2k}$, and the edge joining v_i and v_j is associated with the indeterminate $m_{ij} = -m_{ji}$. If we replace some of the m_{ij} by zeros then we can say that the terms of the Pfaffian correspond to the 1-factors of another graph. That (strict) graph is formed from the complete graph by deleting the edges corresponding to the replaced m_{ij}.

I hoped at first to get a formula for the number of 1-factors of G, analogous to the formula for the number of spanning trees. It is true that by putting $m_{ij} = 1$ or -1 in the non-zero cases we can make each term of P take the value 1 or -1. But I did not see how to get the terms all of the same sign, and my hope was disappointed. Fifteen years later P.J. Kasteleyn circumvented the difficulty in the case of planar graphs [40].

Having got good results from Jacobi's Theorem in the case of the Kirchhoff matrix I was encouraged to apply it to the skew-symmetric matrix M. I then got the following quadratic identity involving Pfaffians:

$$P \cdot P_{qrst} = \pm P_{qr} P_{st} \pm P_{qs} P_{rt} \pm p_{qt} P_{rs}. \qquad (3.3)$$

Here P is the Pfaffian associated with G and P_{qr} is the Pfaffian of the graph G_{qr} obtained from G by deleting the vertices v_q and v_r together with their incident edges. The definitions of P_{qs}, P_{qt} and P_{qrst} are analogous, four vertices and all their incident edges being omitted in the last case. The Pfaffians arose as square roots of determinants. I was therefore left with some ambiguities of sign, but I found I did not need to resolve them. For a resolution, see P.M. Gibson, [29] and [30].

It was convenient to refer to a graph without 1-factors as "prime". Accordingly prime graphs are those for which the associated Pfaffian P vanishes. We can deduce a simple theorem about prime graphs and subgraphs from equation (3.3). Suppose G, G_{qr} and G_{rs} are all prime, then, by (3.3)

$$P_{qs} P_{rt} = 0. \qquad (3.4)$$

There are two possibilities. Either $P_{qs} = 0$ or the Pfaffian P_{rt} is zero for every vertex v_t of G distinct from v_r. In the latter alternative we call v_r a singular vertex or a singularity of G. If v_r is not singular the former alternative holds and we may speak of the transitivity of primality across the non-singular vertex v_r. Accordingly either G_{qs} is prime or G_{rt} is prime for every vertex v_t of G distinct from v_r.

One school of thought holds that theorems on graphs should be proved by graph theory, not by algebra. Hence F.G. Maunsell, after I had published the above proof, wrote a short paper to show that the theorem could be obtained by the purely graph-theoretical method of alternating paths [45]. But however you prove it you can get from it a necessary and sufficient condition for the existence of a 1-factor in a given graph G.

First we introduce some notation. Let S be any set of vertices of G. Then we write G_S for the graph obtained from G by deleting the vertices of S and their incident edges. We call a graph "even" or "odd" according as the number of its vertices is even or odd. We write $h(G, S)$ for the number of odd components of G_S. We denote the cardinality of a given set T by $|T|$.

We can now state the "1-Factor Theorem".

Theorem 3.1 *A graph G is without a 1-factor if and only if it has a set S of vertices such that*

$$h(G, S) > |S|. \tag{3.5}$$

Perhaps I should now state explicitly that in this book I deal only with finite graphs. In discussions of the above theorem we can make a further restriction to strict graphs without losing real generality. For firstly the addition or deletion of a loop to G makes no difference to $|S|$, to $h(G, S)$ or to the 1-factors of G. Secondly the doubling of an edge does not affect $|S|$ or $h(G, S)$ and it cannot introduce a 1-factor into a graph that does not have one already.

For a graph with an odd number of vertices the theorem is trivial. For then there is no 1-factor and (3.5) holds with S null.

In view of these observations we need only prove the theorem for a strict graph G with an even number of vertices. Suppose G has a set S of vertices satisfying (3.5). Then if F is a 1-factor of G and K any odd component of G_S there must be an edge of F with one end in K and one in S. Hence there must be at least $h(G, S)$ edges of F with ends in S, which is impossible. We conclude that G has no 1-factor.

It remains only to show that if G is prime there is an S satisfying (3.5). Let us enumerate the vertices of G as $v_1, v_2, \ldots, v_{2k}$ and define the subgraphs G_{rs} as before.

Let A be an edge of the prime graph G, with ends v_i and v_j. Then G_{ij} is prime, for any 1-factor of it could be made into a 1-factor of G by adjoining the edge A. On the other hand if v_r and v_s are distinct non-adjacent vertices such that G_{rs} is prime, then we can get a new strict prime graph G' from G by adjoining a new edge B with ends v_r and v_s. This is because a 1-factor of

G' not including B would be a 1-factor of G, while one including B would induce a 1-factor of G_{rs}. It follows that by adjoining some edges, possibly none, to G we can transform it into a strict "hyperprime" graph H, that is, a strict graph in which two vertices v_r and v_s are adjacent if and only if H_{rs} is prime.

Let v_q, v_r and v_s be three distinct vertices of G such that, in H, v_q is adjacent to v_r and v_r to v_s. Let v_t be any other vertex of G. Then the three graphs H, H_{qr} and H_{rs} are prime, and therefore equation (3.4) holds in H.

Let S be the set of singularities of H. Each singularity is joined to every other vertex of H by an edge, by its definition and the hyperprimality of H. Each vertex of a component K of H_S is joined to every other vertex of K, by the transitivity of primality. We can deduce from this that

$$h(S) > |S| \qquad (3.6)$$

in H. For suppose not. Then we can find a set of $h(S)$ non-adjacent edges such that each has one end in S and one in an odd component of H_S, and moreover such that each odd component of H_S contains an end of just one of them. To this we can add sets of edges pairing off the remaining vertices of the odd components, then sets of edges pairing off the vertices of the even components, and finally a set of edges pairing off the remaining singularities. For the number of vertices of G, and therefore of H, is even. The result is an impossibility: a 1-factor of H.

Now each odd component of H_S must contain an odd component of G_S. Hence S satisfies (3.5) in G. This completes the proof of the 1-Factor Theorem.

I published this theorem in 1947 [64]. I included a derivation of Petersen's Theorem from the 1-Factor Theorem, a derivation that went essentially as follows.

Let G be a connected cubic graph. Necessarily the number of its vertices is even. Suppose G to be prime. Then it has a set S of vertices satisfying (3.5). Since $h(G, S)$ and $|S|$ have equal parity, by the evenness of G, we can write

$$h(G, S) \geq |S| + 2. \qquad (3.7)$$

Let m be the number of edges joining S to odd components of G_S. Let there be p odd components of G_S joined to S by exactly one edge, and q others. Each of these q is joined to S by at least 3 edges. Since G is cubic we infer that

$$3|S| \geq m \geq p + 3q. \qquad (3.8)$$

Hence, by (3.7),

$$3p + 3q = 3h(G, S) \geq 3|S| + 6 \geq p + 3q + 6, \qquad (3.9)$$

and so $p \geq 3$. The corresponding three or more odd components of G_S would be called "leaves" by Petersen. I note that the edges that join them to S are isthmuses of G not lying on any one arc. So a connected cubic graph has a 1-factor if it has no three isthmuses so related, and that is the required result.

I ought to mention Hall's Theorem, stating a necessary and sufficient condition for the existence of a 1-factor in a bipartite graph. Let me remind you that a bipartition $\{U, V\}$ of a graph G is a partition of its vertex-set $V(G)$ into two complementary subsets U and V such that each edge has one end in U and one in V, and that a bipartite graph is a graph with a bipartition. Evidently a bipartite graph has no loop. Moreover it can have a 1-factor only if

$$|U| = |V| \qquad (3.10)$$

for any bipartition $\{U, V\}$. Suppose condition (3.10) is satisfied. Then Hall's Theorem asserts that G is without a 1-factor if and only if U has a subset U_1 that is joined to fewer than $|U_1|$ vertices of V. The theorem was published by P. Hall in 1934 [35] and it is usually proved by a very simple argument involving alternating paths.

Hall's Theorem may have helped suggest the 1-Factor Theorem to me. It has a similar form but it replaces the "odd components of G_S" by single vertices. Its proof is much shorter than that of Petersen's Theorem, so it seems odd that Petersen's Theorem should have preceded it by 40 years. It can be deduced from the 1-Factor Theorem but T. Gallai thought it more natural to deduce the 1-Factor Theorem from Hall's Theorem [28].

While meditating, at the University of Toronto, on the possibility of a more constructive proof of the 1-Factor Theorem I bethought me of the proof of Petersen's Theorem in Sainte-Laguë's memoir. I set out to write it in a form applying to 1-factors of general graphs, not just cubic ones. When I had got a little way into this project it occurred to me that the argument should work almost as well with f-factors. The upshot was a paper giving a necessary and sufficient condition for the existence of an f-factor in a graph G, published in the *Canadian Journal* in 1952 [65]. I proceed to a statement of this result, beginning with the necessary terminology.

Let us first note that, just as in the theory of 1-factors, there is a trivial parity condition. For an f-factor to exist in G it is necessary that

$$\sum_v f(v) \equiv 0 \pmod{2}. \qquad (3.11)$$

We write $\mathrm{val}(G, v)$ for the valency of a vertex v in a graph G. An f-factor is possible only if

$$\mathrm{val}(G, v) \geq f(v) \qquad (3.12)$$

for each vertex v, so we assume that this condition holds. It is then convenient to write
$$f'(v) = \text{val}(G, v) - f(v) \tag{3.13}$$
for each vertex v. Then there is a one-to-one correspondence between the f-factors and the f'-factors of G, the edges not belonging to a given f-factor defining the corresponding f'-factor.

Consider an ordered partition $\{S, T, U\}$ of the vertex-set $V(G)$ of G into three mutually disjoint subsets. The subgraph of G induced by U—call it $G(U)$—has its set of components. These we classify as "even" and "odd", but not by a rule so simple as in the theory of 1-factors. We now sum $f(v)$ over all the vertices of a component K of $G(U)$, and we add the number of edges joining K to T. We then call K even or odd according as the resulting number is even or odd. We write $h(U)$ for the number of odd components of $G(U)$.

Next we define the "deficiency" $\delta(B)$ of the ordered triad $B = \{S, T, U\}$ as follows:
$$\delta(B) = h(U) - \sum_S f(v) - \sum_T f'(v) + \lambda(S, T), \tag{3.14}$$
where $\lambda(S, T)$ is the number of edges joining S to T. We call B an "f-barrier" of G if
$$\delta(B) > 0. \tag{3.15}$$
The f-Factor Theorem asserts that G has either an f-factor or an f-barrier, but not both.

The f-Factor Theorem is easily extended to cases in which condition (3.11) or (3.12) is violated. Thus if (3.11) fails there is no f-factor and the triad $(\phi, \phi, V(G))$ is an f-barrier. If (3.12) fails there is a vertex w whose valency is less than $f(w)$. Then G can have no f-factor. But $f'(w)$ is negative and therefore $(\phi, \{w\}, V(G) - \{w\})$ is an f-barrier.

Assuming (3.11) and (3.12) it can be shown that $B = (S, T, U)$ is an f-barrier if and only if $B' = (T, S, U)$ is an f'-barrier, and that the number $h(U)$ is the same for B' and f' as for B and f. So the theorem tells us, as we would have expected, that G has an f'-factor if and only if it has an f-factor.

My attempts to apply the f-Factor Theorem gave disappointing results at first. Even the 1-Factor Theorem did not seem to follow from it in any reasonably simple way. On the other hand I was able to deduce the f-Factor Theorem from the 1-Factor Theorem [66]. I did not resolve this difficulty until I had auxiliary theorems saying that under certain conditions one f-barrier could be transformed into another by transferring a vertex from one of the sets S, T and U to another. The auxiliary theorems were published in *Discrete Mathematics* in 1974 [67]. Here I content myself with giving one example of how they work.

Suppose we have an f-barrier $B = (S, T, U)$ such that one vertex w of T satisfies $f(w) = 1$. Let us consider the triad

$$B_1 = (S, T - \{w\}, U \cup \{w\}),$$

formed from B by transferring w from T to U. We study the effect of this transfer on each term on the right of (3.14). We use the information, deducible from (3.11), that $\delta(C)$ is even for every triad C. So our transfer does not alter the parity of $\delta(B)$.

Let p be the number of odd components of $G(U)$ that are joined to w. Then these odd components are combined with w, and perhaps some even components of $G(U)$, to form a single new component of $G(U \cup \{w\})$. So $h(U)$ increases by $\epsilon - p$, where ϵ is 1 or 0 according as this new component is odd or even with respect to B_1.

The sum

$$-\sum_S f(v) - \sum_T f'(v)$$

increases by $f'(w)$, that is by $\text{val}(G, w) - 1$.

The term $\lambda(S, T)$ increases by $-\lambda(S, w)$, that is, minus the number of edges joining w to S. Summing these increases we find that

$$\delta(B_1) - \delta(B) = -p + \epsilon + \text{val}(G, w) - \lambda(S, w) - 1. \qquad (3.16)$$

Note however that $\text{val}(G, w)$ is at least equal to $p + \lambda(S, w)$. Hence $\delta(B_1) - \delta(B)$ is at least $\epsilon - 1$, and therefore at least 0 by conservation of parity. So if B is an f-barrier, then so is B_1.

One consequence of this result is that when we apply the f-Factor Theorem to the case in which $f(v) = 1$ for every v we can deduce that G has either a 1-factor or an f-barrier whose second member T is null. But that is the 1-Factor Theorem.

In that paper of 1974, I show that the f-Factor Theorem, with its auxiliaries, can be used to prove several results of Graph Theory that have emerged since 1947. There is for example the Erdös–Gallai Theorem [25], giving a condition for the existence of a strict graph having a given set of valencies. It can be proved by applying the f-Factor Theorem, with an appropriate function f, to a complete graph.

I feel I ought not to conclude without some further mention of the method of alternating paths. Petersen used this method, Sainte-Laguë used it, Hall used it. Everybody used it save myself in my paper of 1947. The method is based on the theory of a graph whose edges are coloured arbitrarily black and white. The basic problem of that theory is as follows: under what conditions can we get from a given vertex v to some member w of a set

W of other vertices by following a path whose edges are alternately white and black, and which begins with a white edge? Usually the path is allowed to repeat vertices but not edges. Given such a path we can interchange the colours black and white within it, and so get a new edge-colouring of the graph in which no vertex other than v and w has its black valency altered, but in which the black valency of v increases by 1. By repeated application of this interchange operation we may hope to construct an f-factor, defined by black edges, from an initial trivial colouring with no black edges at all. At each stage we can take W to be the set of vertices w whose black valency is less than $f(w)$, and we can assume that no vertex u has black valency exceeding $f(u)$. The construction may of course fail, with a case in which all white-beginning alternating paths from v peter out before getting to W. By considering the set Q of all such paths we can recognize three kinds of edge in the graph G. There are "bicursal edges". Each of these is traversed in each direction by members of Q. The bicursal edges define a subgraph of G whose components, the "bicursal components", are important in the theory. There are "unicursal edges". Each of these is traversed in some direction by some member of Q, but no member of Q traverses it in the other direction. On each unicursal edge Q imposes a definite direction. Finally there are the "acursal" edges, not traversed by any member of Q. Theorems can be proved imposing restrictions on the pattern of the three kinds of edge. For example at most one unicursal edge can enter a bicursal component, though any number can leave it. With these restrictive theorems the pattern resolves itself into an f-barrier. For details see [65]. For an analysis aimed at the construction of algorithms of practical utility see [24].

Why not go a little further and seek conditions for an f-factor satisfying some extra requirement? Why not ask for a connected f-factor, for example? But it seems that this modest request is the proverbial "last straw". It brings us up to an entirely new level of difficulty. Let us remember that a connected 2-factor would be a Hamiltonian circuit. The problem of finding Hamiltonian circuits, generalized as the Travelling Salesman Problem, is of the intractable kind now called "NP-complete".

4

UNSYMMETRICAL ELECTRICITY

The first part of this book described some work on squared rectangles by four undergraduates at Cambridge, who published it in the *Duke Mathematical Journal* in 1940 [11]. At the end of their paper there is a remark on "Triangulations of a triangle". It is about dissections of an equilateral triangle into equilateral triangles. It says there is a theory of such figures analogous to that of squared rectangles, but it does not go into details. There is more information in later papers, for example in one published by the four researchers in the *Philips Research Reports* of 1975. That one is entitled "Leaky electricity and triangulated triangles" [12].

I tell now of dissections of parallelograms, as well as triangles, into equilateral triangles. It is the dissected parallelogram that I find to be the natural analogue of the squared rectangle. Indeed a triangulated parallelogram can be constructed from any squared rectangle by the method presented in Figure 4.1.

We start with the perfect rectangle of Figure 1.1. We "shear it to the right". That is, we move the upper side of the rectangle horizontally to the right with respect to the lower side, and we lower it somewhat, so that each constituent square is transformed into a rhombus with angles of 60

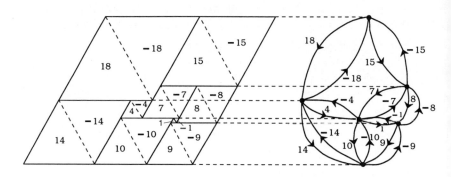

Figure 4.1 A sheared perfect rectangle.

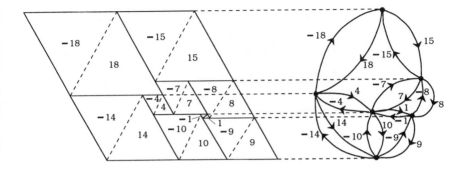

Figure 4.2 The other shearing.

and 120 degrees. Each rhombus is dissected into two equilateral triangles by its short diagonal, shown by a broken line in the figure. Now on the left of Figure 4.1 we have a triangulated parallelogram. On the right we have a complication of the electrical diagram of Figure 1.2, to be explained shortly.

In the dissection of Figure 4.1, and in any triangulated parallelogram, it is clear that one side of each constituent triangle is horizontal, parallel to the lower side of the parallelogram. But the triangle has two possible orientations; it may lie above or below its horizontal side. In the first case we count the side-length of the triangle as positive and in the second negative. The number entered inside each triangle of Figure 4.1 is the side-length, recorded as positive or negative according to the above convention.

Taking advantage of the differences of sign we can say that the 18 constituent triangles of the dissection are all of different sizes. We express this fact by saying that the triangulation is "perfect". We can now state our first theorem: every perfect rectangle shears into a perfect parallelogram. It seems convenient to say that the rectangle and the parallelogram have the same order. So we define the "order" of a triangulated parallelogram as one half of the number of constituent triangles.

In Figure 4.1 the perfect rectangle is sheared to the right. It could equally well have been sheared to the left. That operation gives a different perfect triangulation with the same constituent triangles, and of course with the same horizontal and slanting side-lengths. It is shown in Figure 4.2.

There are some perfect triangulations of parallelograms that are not obtainable by shearing perfect rectangles. I once tried to find the simplest

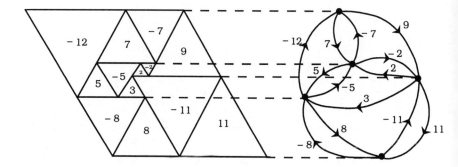

Figure 4.3 A triangulated parallelogram.

such figure and decided that it must be the structure shown above as Figure 4.3. The order of this triangulation is 13/2.

I remember Stone demonstrating, at a meeting of the Four, that in any dissection of an equilateral triangle into smaller ones there must be two constituent triangles with the same (absolute) side-length. The same is true for dissections of parallelograms. Perhaps we should say that a dissected parallelogram can be perfect, but not absolutely perfect. Stone got his result by applying the Euler Polyhedron Formula to the "electrical network" of a triangulated triangle, the network that I have drawn for parallelograms in Figures 4.1, 4.2 and 4.3, and must now explain.

Each horizontal segment in a triangulated parallelogram is represented by a vertex of the "electrical network", just as in the case of a squared rectangle. In Figure 4.3 the vertex is shown in the line of the horizontal segment, but this is not essential. Each constituent triangle is enclosed between two horizontal segments, its base being in one and its apex in the other. It is represented in the network by an edge joining the two corresponding vertices. But this time it is a directed edge, the arrow going from apex to base. Against each directed edge is written the side-length of the corresponding triangle, positive or negative. Let us take over some electrical terminology and say that the upper and lower sides of the parallelogram correspond to the positive and negative poles of the network respectively.

There is a natural anticlockwise cyclic order of the triangles meeting a given horizontal segment by base or apex. We exclude triangles meeting the segment in any other way, of which there are at most two, one at each end. In this cyclic order triangles meeting by base alternate with those meeting by apex. We contrive to preserve this cyclic order in the directed graph. Then at each vertex of the network, incoming edges alternate with outgoing ones. All this is done without edge-crossings, and so our diagram determines a planar map. Let us call it an "alternating map" to emphasize the alternation of incoming and outgoing edges at each vertex. Let us also record the obvious fact that the number of edges at each vertex is even.

The length of the upper horizontal side of a triangulated parallelogram is equal to minus the sum of the side-lengths of the constituent triangles meeting it by base, these side-lengths being all negative. The length of the lower side is the sum of the side-lengths of the triangles meeting it by base. In Figure 4.3 the common value of these two lengths is 19.

The length of any other horizontal segment is the sum of the side-lengths of the triangles meeting it by base and extending above it, that is, the sum for the positive triangles meeting it by base. Similarly the segment-length is minus the sum of the side-lengths of the triangles meeting it by base and extending below it. Combining these results we can say that the sum of the sizes of the constituent triangles meeting the segment by base is zero.

Let us state these results in terms of the alternating map. Let us refer to the number written against each edge as the "current" in that directed edge. Let us define the "current leaving the network" at a given vertex to be the sum of the currents in the incoming edges there. The "current entering the network" at the vertex can be the negative of this. Then the horizontal side of our parallelogram is equal to the current entering at the positive pole and to the current leaving at the negative pole. But at any other vertex the current entering or leaving the network there is zero, that is, the sum of the currents in the incoming edges is zero.

The last observation reminds us of Kirchhoff's First Law, as applied to networks of unit resistances. But now the zero sum is not over all the incident edges but only over the incoming ones.

Kirchhoff's Second Law persists unaltered. We assign to each vertex a "potential" equal to its distance, measured on a slant of 60°, above the base of the parallelogram. Then the current in an edge is the potential at the tail of its arrow minus the potential at the head. It follows that the total change of potential around any circuit is zero, as required by the Second Law.

It seems natural to say that in our network there flows a kind of electricity, with a modified First Law. Smith, using a convention different from the one adopted here, thought of the currents as flowing from triangular base to apex. He said the current was large at the base but it leaked away until finally zero current was delivered at the apex. Hence his name "leaky electricity".

Let us now take note of a procedure for constructing a triangulated parallelogram. First we draw an alternating map. Then we choose positive and negative poles, on the boundary of a specified face called the "outer face". Next we calculate the currents in the wires, or rather their ratios, by solving the modified Kirchhoff equations. Finally we reverse the construction of Figure 4.3.

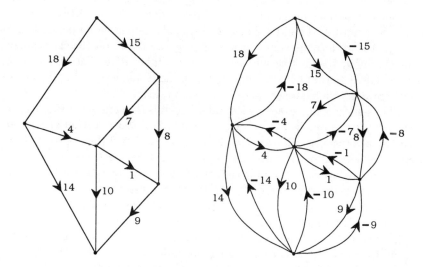

Figure 4.4 Two kinds of electrical networks.

The theory is closely analogous to that of squared rectangles. Indeed we can say more; it is a generalization of that theory. To see this we go back to Figures 1.2 and 4.1. In Figure 4.4 we see on the left the ordinary electrical network of the 32 × 33 squared rectangle, and on the right we see the new-style electrical network of one of its shear-derived triangulated parallelograms.

The main change in going from left to right is that each wire is doubled; it is replaced by two oppositely directed edges. The ordinary Kirchhoff equations for the first network are identical with the modified ones for the second. So the device of doubling edges exhibits the theory of squared rectangles as a subcase of the theory of triangulated parallelograms.

The doubling and directing must be done so as to give an alternating map. Actually this can be done in two ways, corresponding to left and right shearings of the rectangle. As an example we can take the new-style electrical networks appearing in Figures 4.1 and 4.2. The change from the first diagram to the second is simply that the direction of each edge is reversed. It is as a consequence of this reversal that each current is multiplied by -1.

More generally we can observe that the reversal of all edges changes one alternating map into another. If the operation is applied to the electrical network of a triangulated parallelogram it presumably gives the electrical network of another triangulated parallelogram. What is the relation between the two parallelograms? In one special case we know that they

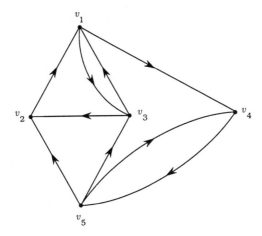

Figure 4.5 An "unsymmetrical" electrical network.

are right and left shearings of the same squared rectangle. In the case of Figure 4.3 the network and its parallelogram are merely replaced by their mirror images. More general cases will be discussed later, in connection with modified electrical network theory.

Let us discuss the modified Kirchhoff equations. We will apply them to a general graph G with directed edges. This graph need not be planar and it need not define an alternating map. We do not require the number of incoming edges at a vertex to be equal to the number of outgoing ones. But if this equality does hold at every vertex we shall say that G is "balanced".

We define the Kirchhoff matrix $K(G)$ of G much as before. Let the vertices be enumerated as $v_1, v_2, \ldots, v_n$. Then $K(G)$ is a square matrix of order n. We define the jth diagonal entry c_{jj} as the number of darts (directed edges) coming to v_j from other vertices. The non-diagonal entry c_{ij} in the ith row and jth column is minus the number of darts directed from v_j to v_i. Suppose for example that G is the directed graph shown in Figure 4.5.

Then the matrix $K(G)$ is as follows.

$$\left\{ \begin{array}{rrrrr} 2 & -1 & -1 & 0 & 0 \\ 0 & 2 & -1 & 0 & -1 \\ -1 & 0 & 2 & 0 & -1 \\ -1 & 0 & 0 & 2 & -1 \\ 0 & 0 & 0 & -1 & 1 \end{array} \right\}$$

We note that in this case $K(G)$ is not symmetrical. A better name for "leaky electricity" might be "unsymmetrical electricity". I propose to use the latter name from now on.

We note also that the elements of $K(G)$ sum to zero in the rows but not in every column. However this is enough to ensure that

$$\det K(G) = 0, \tag{4.1}$$

just as in the symmetrical case.

We define the "complexity" $C_i(G)$ of G at v_i as $\det K_i(G)$, where $K_i(G)$ is the matrix obtained from $K(G)$ by striking out the ith row and column. In the symmetrical case we found that determinant to be independent of i. But the proof of that independence uses the fact that the elements of $K(G)$ sum to zero in the columns as well as in the rows, and that need not be true in the unsymmetrical case. It is true for balanced directed graphs but not for unbalanced ones. However the alternating maps of the theory of triangulated parallelograms do correspond to balanced graphs, and for these the complexity is the same at every vertex, just as in the symmetrical case. We can speak without ambiguity of "the complexity" of an alternating map.

The graph of Figure 4.5 is not balanced, and it has a different complexity at each vertex. We calculate $C_1(G)$ as follows:

$$C_1(G) = \begin{vmatrix} 2 & -1 & 0 & -1 \\ 0 & 2 & 0 & -1 \\ 0 & 0 & 2 & -1 \\ 0 & 0 & -1 & 1 \end{vmatrix} = 4 \begin{vmatrix} 2 & -1 \\ -1 & 1 \end{vmatrix} = 4.$$

The other complexities can be calculated similarly. It can be verified that $C_2(G) = 2, C_3(G) = 3, C_4(G) = 5$ and $C_5(G) = 10$.

The usual form of the Matrix–Tree Theorem says that the complexity of an undirected graph is equal to the number of its spanning trees. There is a generalized form of the theorem applying to directed graphs. It says that $C_i(G)$ is the number of spanning trees of G diverging from v_i. The phrase "diverging from v_i" means that each edge is directed away from v_i in the tree. When an edge A is deleted from a spanning tree, the tree falls into two pieces. We require the arrow of A to be directed from the part containing v_i to the part not containing it. I published this general form of the theorem in 1948 [68]

Since $C_2(G) = 2$ the generalized Matrix–Tree Theorem asserts that G has just two spanning trees diverging from v_2. These two trees are shown in Figure 4.6.

In a balanced graph the complexity is the same at every vertex. In such a graph therefore the same number of spanning trees must diverge from every vertex.

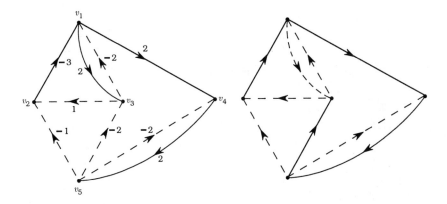

Figure 4.6 Directed trees.

Now let us discuss a distribution of current satisfying the modified Laws. Let us take the first vertex v_1 to be the positive pole and the last vertex v_n to be the negative pole. Let us write V_1 for the potential of v_1 in a corresponding flow. Writing the current in each wire as a difference of potentials we find that the current entering the network at any vertex v_i is given by

$$\sum_j c_{ij} V_j. \tag{4.2}$$

This we equate to zero if i is not 1 or n. For $i = 1$ or n we write it as U or $-W$ respectively, where we expect U and W to be non-negative. Let us take V_n to be zero, and let us ignore the nth equation. We then have a system of linear equations with the matrix $K_n(G)$, and they have a unique solution by Cramer's Rule whenever $C_n(G)$ is non-zero. Each potential, other than V_n, is then expressed as a signed minor of $K(G)$ multiplied by $U/C_n(G)$. We can therefore make all the currents take integral values by adopting the convention that $U = C_n(G)$. We refer to the resulting flow as the "full flow" from v_1 to v_n. As in the theory of squared rectangles we refer to the highest common factor of its currents as the "reduction" of the flow. In particular the full potential drop from v_1 to v_n is found to be $\det K_{1n}(G)$, where $K_{1n}(G)$ is the matrix derived from $K(G)$ by striking out the first and last rows and columns. In this full flow the current leaving the network at the negative pole is found to be $C_1(G)$.

In the first diagram of Figure 4.6 I have shown the reduced flow from v_1 to v_5 (in the notation of Figure 4.5). The current entering at the positive

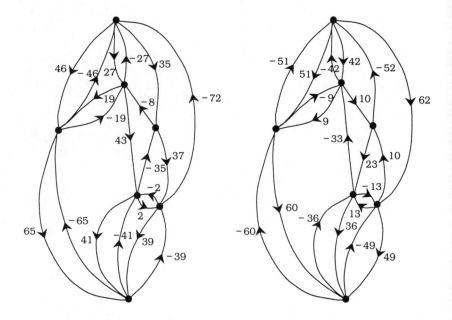

Figure 4.7 Nets of two parallelograms of the same shape.

pole is 5 and that leaving at the negative is 2. The reduction is 2. For unbalanced graphs those two currents do not have to be equal.

Expositions in greater detail of the theory of unsymmetrical flows can be found in [12], [68] and [71].

Let us return to the operation of reversing every dart in a network. For a balanced graph, and for a balanced graph only, it replaces $K(G)$ by its transpose. It then likewise replaces $K_1(G)$ and $K_{1n}(G)$ by their transposes, and so does not alter their determinants. Hence if G is the graph of an alternating map the operation alters neither the horizontal side nor the slanting side of a corresponding triangulated parallelogram. Nevertheless it may change the set of full currents in the wires of the network, and therefore it may alter the set of constituent triangles in the parallelogram. It may even happen that the two triangulations have no constituent triangle-size in common. An example is given in Figure 4.7.

The electrical networks are shown here, with their currents, but the parallelograms are omitted. Their order is 10. Another such pair is drawn in full in [70]. The similar pair in [68] has a repeated constituent of size 129.

Is there any analogue of duality in the theory of triangulated parallelograms? The answer is "Yes". But it is best explained in terms of triangulated triangles rather than triangulated parallelograms. It is easy to

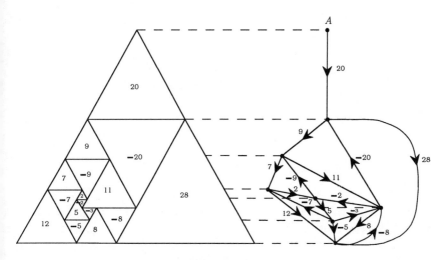

Figure 4.8 A triangulated triangle.

convert a triangulated triangle into a triangulated parallelogram, or vice versa, by adding one or two extra constituent triangles. In Figure 4.8 I give an example, due to L. Augusteijn, with its electrical network. The network shows the horizontal segments as vertices and the constituent triangles as directed edges, just as for a parallelogram, except that there is an extra vertex A to represent the apex of the triangle. This extra vertex is the positive pole of the network.

The electrical network is not that of an alternating map, but this defect is easily corrected by identifying the two poles. Let us note that the current entering the network at the positive pole is zero. This is in accordance with the Matrix–Tree Theorem, since there can be no spanning tree diverging from any vertex other than the positive pole, and therefore the complexity at the negative pole is zero.

Figure 4.9 shows the electrical network obtained from that of Figure 4.8 by identifying the two poles. The edge incident with the positive pole A in Figure 4.8 we call the "polar edge" of the new diagram and mark with a cross. If we assign it a current -28 we get the electrical network of a parallelogram, complete with currents. That parallelogram is obtained from the triangulated triangle by suppressing the constituent of size 20 and adding a new one of size -28 on the right. It can be shown that the

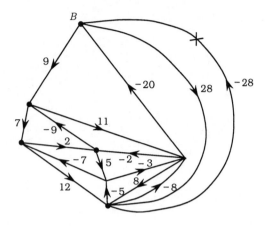

Figure 4.9 From triangle to parallelogram.

complexity of the new network is equal to that of the original at A. To do this by the Matrix–Tree Theorem we observe that the spanning trees diverging from B in Figure 4.9 become the spanning trees diverging from A in Figure 4.8 when the edge AB is adjoined.

Now the triangulated triangle of Figure 4.8 has three sets of parallel segments, and any one of the three kinds can be chosen as "horizontal". Using each set in turn we can construct three different electrical networks, which become three different alternating maps when poles are identified. We may see this procedure as analogous to the construction of two dual c-nets from the two sets of parallel segments in a squared rectangle. But three alternating maps are not to be called duals. I refer to them as "trine"[1] maps. The property they exhibit, analogous to duality, is called "triality" in the 1940 paper, and "trinity" in some later ones ([69], [3]).

Working out how best to fit three trine maps together in one diagram is a fascinating exercise, but a full account of how the Four arrived at a solution might be rather lengthy. Let me go directly to Smith's account of trinity in terms of a planar bicubic map.

A "bicubic" map is a map that is both cubic and bipartite. P.J. Heawood showed that a cubic map can be face-coloured in three colours if and only if it is bicubic. Bicubic maps are aslo well known in connection with Barnette's Conjecture that every 3-connected planar bicubic map has a Hamiltonian circuit.

Figure 4.10 shows a planar bicubic map. The vertices are marked alternately + and −, and the faces are coloured in three colours I, II and III.

[1] A term borrowed from astrology.

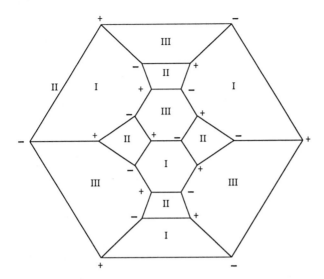

Figure 4.10 A bicubic map.

We can derive an alternating map from this figure as follows. The vertices are the map-faces of colour I. The edges are the map-edges separating the colours II and III. Each of these is directed from its negative end to its positive; it can be regarded as going from one I-face to another. If digons are allowed such an edge may begin and end at the same I-face. The new structure becomes an alternating map when each I-face is contracted to a point. Similar operations give alternating maps with the II-faces, or the III-faces, as vertices. These three alternating maps are found to be the three trine maps of a triangulated triangle. Conversely three trine alternating maps can be related to a bicubic map in the above way [12].

The main theorem about trinity is that three trine alternating maps have equal complexities, just as do two dual c-nets. [Problem for the reader: what happens to the third "dual" when we pass from triangulated parallelograms to their special case of squared rectangles?] The proofs of the theorem in [68] and [69] are rather long. But recently K. Berman replaced them by a short algebraic argument [3].

Smith asked: What is the graph-theoretical interpretation of the complexity as a property of the bicubic map? He decided that it was the number of 1-factors of a certain kind. The question is discussed in the Four's paper of 1975 [12].

5

ALGEBRA IN GRAPH THEORY

I learned a little Combinatorial Topology at Cambridge. That subject dealt with structures called "n-complexes". These were made by fitting together units called "n-simplexes", n-dimensional analogues of the point, the segment, the triangle and the tetrahedron. The graphs of Graph Theory appeared as the case $n = 1$, and by combinatorial topologists they were usually looked upon as trivialities. However O. Veblen devoted part of a textbook to them and to their associated maps on surfaces [108]. The topological textbook then favoured at Cambridge was that of Seifert and Threlfall [52].

The message of these textbooks seemed to be that progress could be made in Combinatorial Topology, and therefore perhaps even in Graph Theory, by adorning one's simplexes with algebraic symbols manipulatable by the methods of Linear Algebra. I soon perceived that the familiar 4-colourings of planar maps, the Tait colourings of cubic graphs, and the congruences mod 3 with which P.J. Heawood was still attacking the Four Colour Problem were special cases of the "n-chains" and "n-cycles" of Combinatorial Topology ([36], [37]).

Consider for example the "cycles mod 2" that are so often put upon a graph. With each edge E of a graph G we can associate a residue $f(E)$ modulo 2. The function f from $E(G)$ to the set I_2 of residues mod 2 is called a "1-chain on G over I_2". It is natural to associate the 1-chain with the subset of $E(G)$ made up of those edges E for which $f(E) = 1$, and even to identify it with this subset.

There is a rule of addition for 1-chains: The sum of two 1-chains f and g on G over I_2 is the 1-chain h on G over I_2 such that

$$h(E) = f(E) + g(E) \tag{5.1}$$

for each edge E of G. To expound the definition properly I should now demonstrate that this addition is commutative and associative and that it can be extended to sums of three or more 1-chains. But let me skip that part. There is a "zero chain" in which the coefficient $f(E)$ is 0 for every E. It is represented in formulae by a simple zero, as in equation (5.2) below, which asserts that every 1-chain is its own negative. Subtraction for 1-chains over I_2 is the same as addition.

$$f + f = 0. \tag{5.2}$$

Likewise there are "0-chains on G over I_2". They are the same as 1-chains except that they are defined on $V(G)$, the set of vertices, instead of $E(G)$. With each 1-chain f we now associate a 0-chain ∂f called the "boundary" of f. The coefficient of a vertex v in ∂f is, by definition, the sum of the coefficients $f(E)$ over all links E incident with v. (A link is an edge with two distinct ends). Let us now pause to note and emphasize the linear identity

$$\partial(f + g) = \partial f + \partial g, \tag{5.3}$$

valid for arbitrary 1-chains f and g. We note also that the boundary of the zero 1-chain is the zero 0-chain: $\partial 0 = 0$.

Sometimes a 1-chain f has the zero 0-chain as its boundary. The 1-chain is then called a "1-cycle", or simply a "cycle" of G. The term "1-cycle" would be insisted upon in more general Combinatorial Topology, which recognizes cycles of dimensions other than 1. From the linear identity (5.3) we can infer the theorem that any sum of cycles is a cycle.

We have a vector space of 1-chains on G over I_2, and within it a vector space of cycles on G over I_2. A cycle f can be identified with a subgraph in which the valency of each vertex is even. The subgraph is defined by the edges E of G for which $f(E) = 1$, and by their incident vertices. In particular the edges of a circuit, or of a union of disjoint circuits, define a cycle. In a cubic graph the only cycles are the zero 1-chain and those corresponding to circuits and unions of disjoint circuits. In these essays the cycle corresponding to a circuit is called an "elementary cycle", and the cycle corresponding to a Hamiltonian circuit is a "Hamiltonian cycle".

We may find this structure impressive but ask "Does it really help us to prove graph-theorems?" The answer is that sometimes it does. Take, for example, Smith's Theorem that in every cubic graph G the number of Hamiltonian circuits passing through any specified edge is even. In our present terminology we can state Smith's Theorem as follows: the Hamiltonian cycles of a cubic graph G sum to zero. We can then prove it as follows.

Consider a Tait colouring T of G. The edges are coloured in three colours α, β and γ so that no two of the same colour meet at a vertex. We do not regard a permutation of the three colours as giving a new Tait colouring. Taking the edges of two colours α and β we get a Tait cycle $T_{\alpha\beta}$. Similarly we have Tait cycles $T_{\beta\gamma}$ and $T_{\gamma\alpha}$. These "Tait cycles" really are cycles; each of them corresponds to a circuit or to a union of disjoint circuits. If a Tait cycle corresponds to a circuit it is a Hamiltonian cycle. In any case each component circuit of a Tait cycle is of even length. Since each edge of G belongs to exactly two Tait cycles of each Tait colouring we have the identity

$$T_{\alpha\beta} + T_{\beta\gamma} + T_{\gamma\alpha} = 0. \tag{5.4}$$

It is valid for every Tait colouring T.

Now let K be any cycle of G corresponding to a union of disjoint circuits of G, all of even length, that spans G. We write k for the number of its component circuits. We take this definition to include the case $k = 1$, the case in which K is Hamiltonian. We can interpret K as the Tait cycle $T_{\alpha\beta}$ of some Tait colouring. We do this by colouring the edges of each constituent circuit alternately α and β, and then giving the colour γ to all the remaining edges of G. We note that this can be done in 2^k ways. However these ways occur in equivalent pairs, the members of each pair differing only by a permutation of the colours α and β. We conclude that the number of distinct Tait colourings having K as a Tait cycle is

$$2^{k-1}.$$

We now sum equation (5.4) over all the distinct Tait colourings of G, finding that

$$\sum_K 2^{k(K)-1} K = 0, \tag{5.5}$$

where the sum is over all possible choices of the cycle K and $k(K)$ is the number of constituent circuits of K. Since our coefficients are residues mod 2, only cycles K for which $k(K) = 1$ make a non-zero contribution to the left side of (5.5). Accordingly (5.5) asserts Smith's Theorem; the Hamiltonian cycles of G sum to zero.

With this encouragement we go on to a more general theory of 0-chains and 1-chains, by a further borrowing from Algebraic Topology. We mark a direction on each edge of our graph G. But now this arrow is not regarded as an essential part of the mathematical structure being studied, as it was in the discussion of unsymmetrical electricity. It is there merely for descriptive convenience. We define 0-chains and 1-chains by assigning to each vertex or directed edge a coefficient from some specified ring R. Thus a 0-chain or 1-chain is a function f from $V(G)$ or $E(G)$ respectively to R. We say that the chain is on $V(G)$ or $E(G)$ and over R.

Addition of 0-chains or 1-chains over R is defined in the obvious way. We have also the operation of forming λf, the product of an element λ of R by a 0-chain or 1-chain f. In a formal development we would now prove that this operation is distributive over addition of chains. The difference $f - g$ of two 0-chains or two 1-chains can now be introduced as $f + (-g)$, that is $f + (-1)g$. We assume here that R has a unit element. In this essay we shall also take R to be commutative.

Commonly R will be the ring I of integers, or the ring I_m of residues modulo some integer m. We have already discussed the case $R = I_2$. In another interesting case R is the 4-ring, with its four elements $0, 1, \omega$ and ω^2. Its rules of addition are

$$x + x = 0, \quad 1 + \omega + \omega^2 = 0. \tag{5.6}$$

One rule of multiplication is $\omega^3 = 1$, and the others are indicated by the notation. If R is I_2 or the 4-ring then each element of R is its own negative.

In this theory the arrow on any edge E of G can be reversed at our pleasure, provided that we also replace the coefficient of E by its negative in every 1-chain. When R is I_2 or the 4-ring, reversals of arrows make no difference to the coefficients, and we may as well work with undirected edges.

The definition of the boundary ∂f of a 1-chain f on G over R now needs a little more care. It is still a 0-chain on G over R. But now the coefficient in ∂f of a vertex v is the sum of the coefficients in f of the edges directed to v minus the corresponding sum in the edges directed from v. Thus a loop would make two contributions, but they would cancel out. If we think of the coefficient $f(E)$ as a current in the directed edge E, then we can say that $(\partial f)(v)$ is the current leaving the network at v. Evidently the sum of the outgoing currents for all the vertices is zero. Let us note that if we reverse some arrows and correspondingly adjust the signs of the coefficients in f, then we do not alter the boundary ∂f. Let us note also that the identity (5.3) is still valid.

It may happen that ∂f is the zero chain. We then call f a 1-cycle, or simply a cycle, on G over R. The set of such cycles is closed under the operations of addition and multiplication by an element of R. To get an example of a cycle we can assign a non-zero coefficient to some loop of G, and zero currents to all the other edges. A Tait colouring can be interpreted as a cycle over the 4-ring with no zero coefficient; we need only take α, β and γ to be $1, \omega$ and ω^2 respectively.

Consider a circuit C in G, and suppose the arrows in its edges to be directed all the same way around C. Giving each dart of C the same coefficient λ from R, and giving every other directed edge of G the coefficient zero, we obtain a corresponding cycle of G. More generally, with the arrows in C unrestricted, we would give the coefficient λ to darts going one way around C and the coefficient $-\lambda$ to those going the other way.

Now let us have an application to Graph Theory. Consider a planar map M. For simplicity we will suppose each face to be simply connected, that is, bounded by a single circuit. The map is on a sphere, and a positive sense of rotation around each face is defined. In any edge the positive senses of rotation in the two incident faces induce opposite directions, opposite darts. By an "m-colouring" of M we mean a colouring of the faces in m colours so that no two faces with a common edge have the same colour. Let G be the graph defined by the edges and vertices of M.

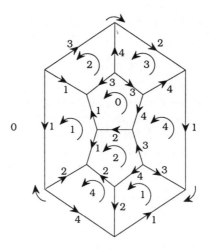

Figure 5.1 A 5-colouring and its bounding cycle.

We identify the m colours with the m elements of I_m. Assigning colours to the faces, each taken with its positive orientation, gives us a 2-chain on M over I_m, in the language of Combinatorial Topology. To any face F of colour α there corresponds a cycle ∂F in which the edges of the bounding circuit have coefficients α or $-\alpha$ according as they are directed in the positive or the negative sense around F, and in which the other edges of G have zero coefficients. Adding the cycles ∂F of all the faces we obtain a cycle g on G over I_m, technically known as the boundary of the 2-chain. It is a "nowhere-zero" cycle; that is, its coefficients are all non-zero. This is because any two faces with a common edge must have different colours. Figure 5.1 shows a 5-colouring and its bounding cycle g. The numbers written in this diagram are to be interpreted as residues mod 5. Those shown centrally in the faces are coefficients in the 2-chain, and those written against the directed edges are coefficients in g.

It is a theorem of Combinatorial Topology that for a map on the sphere every 1-cycle over R bounds a 2-chain over R. Hence every nowhere-zero cycle on G over I_m must bound an m-colouring of M. So the problem of finding an m-colouring for M is equivalent to that of finding a nowhere-zero cycle over I_m for G.

The same argument applies with any ring of coefficients having finitely many elements. In particular it applies with the 4-ring. Hence the Four Colour Problem for cubic maps, without isthmuses, is equivalent to that of showing that each such map has a Tait colouring.

Note how the new theory generalizes face-colouring problems. In a non-planar graph we do not have faces to colour, but we can still search for nowhere-zero cycles over I_m. A planar problem is suddenly generalized to all graphs.

A graph G may have an isthmus. As the reader may like to verify, an isthmus of G must have zero coefficient in every cycle. So a graph with an isthmus has no nowhere-zero cycle over any R. However when it is drawn on the plane it does not give an acceptable map for colouring, for the isthmus has the same face on both sides.

A generalized Four Colour Conjecture, saying that every graph without an isthmus has a nowhere-zero cycle over I_4, would find counter-examples among the non-planar cubic graphs. The Petersen graph would be one. That is because the Petersen graph has no Tait colouring, and because any nowhere-zero cycle g over I_4 in a cubic graph G determines a Tait colouring. To prove the latter statement consider any vertex v of G. One edge there must have the residue 2 as its coefficient in g, and each of the others must have coefficient 1 or -1. Deleting from G all the edges with coefficient 2 we therefore obtain a spanning subgraph K which is a union of disjoint circuits. Let K_v denote whichever of these circuits goes through a given vertex v. If its arrows are adjusted to go all the same way around K_v, then one of the edges incident with v must have coefficient 1 and the other coefficient -1 in the cycle g. Hence K_v must be of even length, the coefficients 1 and -1 alternating around it. Accordingly K is a Tait cycle of G.

The preceding result is a special case of a general theorem. The number of nowhere-zero cycles of a graph G over a finite ring R depends on the number of elements of R, but not otherwise on the nature of R. (See [72]).

The term "snark" has come into use for a 3-connected cubic graph with no Tait colouring, or equivalently with no nowhere-zero cycle over I_4. Many snarks, and even some infinite families of snarks, have been found ([9], [19], [38]).

The Five Colour Theorem asserts that the faces of a planar map can be coloured in 5 colours so that no two of the same colour share a common edge. It has a very simple proof, and textbooks on Graph Theory are apt to dismiss it as one of the most trivial of those theorems that have acquired proper names. But its generalization, asserting that every graph without an isthmus has a nowhere-zero cycle over I_5, cannot be so dismissed. For it still lacks proof or disproof. It is the 5-Flow Conjecture. However the corresponding 8-Flow Conjecture has been proved by F. Jaeger and the 6-Flow Conjecture by P.D. Seymour ([39], [53]).

I use the term m-flow, where m is an integer greater than or equal to 2, to denote a nowhere-zero cycle over I such that each of its coefficients is less than m in absolute value. From an m-flow we can obtain a nowhere-zero cycle over I_m by replacing each coefficient by its residue mod m. The converse result is true, but not trivial: given any nowhere-zero cycle mod m

we can always transform it into an m-flow by replacing each coefficient by an appropriate corresponding integer. There is a proof in [73]. Because of these results it does not matter whether we regard the 5-Flow Conjecture as asserting the existence of a 5-flow or a nowhere-zero cycle over I_5.

The generalized 5-Flow Conjecture, like its planar subcase, can be reduced to a theorem about cubic graphs. The speculation that it may be false implies that some day we may meet a super-snark, a 2-connected cubic graph with no 5-flow. Such a graph would obviously have no 4-flow, and therefore it would be a true snark, though more than a snark.

It is written:

> For though common snarks do no manner of harm
> Yet I feel it my duty to say
> Some are Boojums–The Bellman broke off in alarm,
> For the Baker had fainted away.

Perhaps the Baker is an allegorical figure representing the 5-Flow Conjecture. He tells of an old uncle's warning.

> But O beamish nephew, beware of the day
> If your snark be a Boojum, for then
> You will softly and suddenly vanish away
> And never be met with again. [16]

Let us now consider a second linear operation on chains, that of taking the coboundary δh of a 0-chain h on G over R. Given any edge E of G whose arrow goes from the end u to the end v we assign to it the coefficient $h(u) - h(v)$. We thus obtain a 1-chain on G over R, and this 1-chain is called the coboundary of h. In an electrical analogy, with unit resistances, we can regard the coefficients $h(w)$ as the potentials of the vertices w. Then the coefficients in δh are the currents in the edges, each in the direction of the corresponding arrow. We deduce from this definition that coboundaries satisfy the following linear identity:

$$\delta(g + h) = \delta g + \delta h. \tag{5.7}$$

An m-colouring of a graph G is a colouring of its vertices in m colours so that no edge has both its ends of the same colour. This restriction is held to imply that no graph with a loop can have an m-colouring. Such an m-colouring can be interpreted as a 0-chain h on G over I_m. Evidently the condition that a given 0-chain over I_m shall be an m-colouring of G is that it shall have a nowhere-zero coboundary. So the problem of m-colouring a graph G is that of finding a nowhere-zero coboundary on G over I_m.

The number $P(G, m)$ of m-colourings of a loopless graph G is of great interest in algebraic Graph Theory. It is found to be a polynomial in m of degree $|V(G)|$, and it is therefore called the "chromatic polynomial" or "chromial" of G. If G has n components then its chromial divides by m^n, and $m^{-n}P(G,m)$ is the number of nowhere-zero coboundaries on G over I_m. The number $F(G,m)$ of nowhere-zero cycles on G over I_m is another polynomial that has been much studied. It is called the flow-polynomial.

Let A be a link of the graph G. I often denote the result of deleting A from G by G'_A, and the result of contracting A to a single vertex by G''_A. In their study of complexities or tree-numbers the Four made much use of the recursion formula

$$C(G) = C(G'_A) + C(G''_A). \tag{5.8}$$

When I was doing my PhD research I began to collect other functions of graphs that satisfied similar recursions. In a footnote to [114] I found a theorem of R.M. Foster: chromatic polynomials of graphs satisfy the recursion

$$P(G,m) = P(G'_A, m) - P(G''_A, m). \tag{5.9}$$

I sought something similar for the flow-polynomial and discovered that

$$F(G,m) = F(G''_A, m) - F(G'_A, m). \tag{5.10}$$

These observations led eventually to the paper [72]. In it I discussed a function f of graphs satisfying the rules

$$f(G) = f(G'_A) + f(G''_A), \tag{5.11}$$

$$f(H+K) = f(H)f(K). \tag{5.12}$$

Here A is any link of G, and $H+K$ means a graph which is the union of two disjoint subgraphs H and K. We can deduce from (5.9) and (5.10) that the above two equations hold when f is one of the following functions of graphs:

$$(-1)^{|V(G)|}P(G,m), \quad (-1)^{|V(G)|+|E(G)|}F(G,m).$$

Let us write $p_0(G)$ for the number of components or connected pieces of a graph G. Let us write also $p_1(G)$ for the "cyclomatic number" of G, definable by the equation

$$p_1(G) = |E(G)| - |V(G)| + p_0(G). \tag{5.13}$$

Attempts to find a sum over subgraphs that satisfied (5.11) and (5.12) succeeded with the following "dichromatic polynomial" $Q(G; t, z)$.

$$Q(G;t,z) = \sum_S t^{p_0(G:S)} z^{p_1(G:S)}. \tag{5.14}$$

Here S runs through the subsets of $E(G)$, and $G:S$ denotes the spanning subgraph of G whose edges are the members of S. The polynomial is in two indeterminates t and z.

The dichromatic polynomial can be related to the chromatic polynomial and to the flow-polynomial by the following equations:

$$P(G,m) = (-1)^{|V(G)|}Q(G;-m,-1), \qquad (5.15)$$
$$F(G,m) = (-1)^{|E(G)|+|V(G)|}Q(G;-1,-m). \qquad (5.16)$$

Using (5.14) we can rewrite (5.15) as follows:

$$P(G,m) = \sum_S (-1)^{|S|} m^{p_0(G:S)}. \qquad (5.17)$$

This is Whitney's famous explicit formula [115].

I often used a modification $\chi(G;x,y)$ of the dichromatic polynomial that seemed to me to be simpler and neater. It is defined as follows:

$$\chi(G;x,y) = (x-1)^{-p_0(G)} Q(G; x-1, y-1). \qquad (5.18)$$

It is a polynomial in two variables x and y. In my papers[2] I call it the "dichromate" of G.

The dichromate satisfies (5.11) only when A is an edge of G that is neither a loop nor an isthmus. However it obeys a very helpful recursion formula for graphs that are connected but separable. If a connected graph G is the union of two subgraphs H and K that have only a vertex in common, then

$$\chi(G;x,y) = \chi(H;x,y)\chi(K;x,y). \qquad (5.19)$$

A further pleasing property is that if G and G^* are dual planar connected graphs, then

$$\chi(G;x,y) = \chi(G^*;y,x). \qquad (5.20)$$

Let us now consider how to make a catalogue of dichromates of connected graphs. For very simple graphs we can use the explicit formula (5.14) to calculate Q and then deduce χ. For us these very simple graphs will be the graphs with at most one edge. They include the "vertex-graph", which has a single vertex and no edges. For it $\chi(G;x,y) = 1$. Then there is the "link-graph", consisting of a single edge with two distinct ends. For it we find that $\chi = x$. Last there is the "loop-graph", which has a single edge and a single vertex, the edge being a loop. For the loop-graph $\chi = y$. The link-graph and loop-graph can be taken as G and G^* in (5.20).

Let n be a positive integer. Let B_n be a graph with just two vertices and just n edges, each edge being a link joining the two vertices. I will call B_n the "n-bond". Let C_n denote the circuit of n edges. Evidently B_1 is the link-graph and C_1 the loop-graph. We note also that B_2 and C_2 are

---

[2] It is now usually called the "Tutte polynomial".

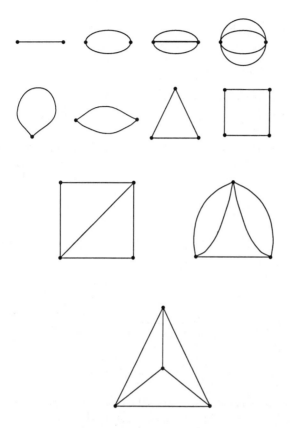

Figure 5.2 Some special graphs.

identical. The first row of Figure 5.2 shows the n-bonds from $n = 1$ to $n = 4$, and the second row the corresponding n-circuits. The third row shows two more complicated graphs. For the time being I will refer to the first as the "crossed 4-circuit" and to the second as the "split 4-bond". On the fourth row we have the complete 4-graph, also called the "4-clique" and the "tetrahedral graph".

Consider an edge A of B_n, where $n \geq 2$. Then A is neither a loop nor an isthmus, G'_A is an $(n-1)$-bond and G''_A is made up of a single vertex and $n-1$ loops. So we can use (5.11) and (5.19) to get the recursion

$$\chi(B_n; x, y) = \chi(B_{n-1}; x, y) + y^{n-1}. \tag{5.21}$$

Now we can enter in our catalogue the dichromates $\chi(B_1) = x$, $\chi(B_2) = x + y$, $\chi(B_3) = x + y + y^2$, and so on.

Next we take A to be an edge of C_n, where $n \geq 2$. We find that G'_A is an arc of $n-1$ edges and that G''_A is an $(n-1)$-circuit. Hence

$$\chi(C_n; x, y) = x^{n-1} + \chi(C_{n-1}; x, y). \tag{5.22}$$

We can now make the entries $\chi(C_1) = y$, $\chi(C_2) = x+y$, $\chi(C_3) = x^2+x+y$, and so on. But we could have shortened the work by observing that C_n is the planar dual of B_n, and then using (5.20).

Next let G be the crossed 4-circuit of Figure 5.2, and let us take A to be the crossing or diagonal edge. Then G'_A is C_4. But G''_A is a separable graph, consisting of two copies of C_2 with only a vertex in common. We infer that

$$\begin{aligned}\chi(G) &= (x^3 + x^2 + x + y) + (x+y)^2 \\ &= x^3 + 2x^2 + 2xy + y^2 + x + y.\end{aligned} \tag{5.23}$$

Now take G to be the split 4-bond. Since that graph is the planar dual of the crossed 4-circuit we can write at once

$$\chi(G) = y^3 + 2y^2 + 2xy + x^2 + x + y. \tag{5.24}$$

Last, we take G to be the 4-clique. With any edge as A we find that G'_A is the crossed 4-circuit and G''_A is the split 4-bond. So we can get $\chi(G)$ by adding the right sides of (5.23) and (5.24):

$$\chi(G) = x^3 + y^3 + 3x^2 + 4xy + 3y^2 + 2x + 2y. \tag{5.25}$$

At this stage in the construction of the catalogue it becomes convenient to represent χ by a table of coefficients, putting that of $x^{i-1}y^{j-1}$ in the ith row and jth column, and omitting zeros when convenient. Thus the table for the crossed 4-circuit is as follows:

$$\begin{pmatrix} 0 & 1 & 2 & 1 \\ 1 & 2 & & \\ 1 & & & \end{pmatrix},$$

and that for the 4-clique is

$$\begin{pmatrix} 0 & 2 & 3 & 1 \\ 2 & 4 & & \\ 3 & & & \\ 1 & & & \end{pmatrix}.$$

The symmetry of this table could have been predicted from the fact that the 4-clique, like the 2-circuit, is a self-dual graph.

One can go on to make quite an extensive catalogue. G. Berman has computerized the procedure. Inspection of a catalogue shows that the entries in each table are all non-negative. But there is no mystery about that;

it follows obviously from the recursion formulae. Another observation is that every row and column, in every table, has the unimodal property: no entry is less than both its neighbours. That property is still a mystery; no one has yet proved it as a theorem or found a counter-example.[3] The same unimodal property holds for the sequence of coefficients of the powers of m in $P(G,m)$, or rather for their absolute values, with the same lack of explanation.

When I first began to calculate dichromates I was much intrigued by the fact that the sum of the coefficients in $\chi(G;x,y)$, for a connected graph G, was equal to the complexity $C(G)$ of G. That result is easily proved by induction, using known recursions. It suggested to me that the dichromate should be expressible as a sum over spanning trees. But that seemed impossible in the case of the 3-circuit. Surely the three spanning trees would have to give the three distinct terms x^2, x and y of the dichromate. And yet these trees were all alike under the symmetry of the graph.

The way out of this difficulty was to impose an arbitrary enumeration on the edges of the graph. Then for each tree T numbers $r(T)$ and $s(T)$ could be defined which expressed the relation of T to the fixed enumeration of edges. That enumeration of course destroyed all symmetry. It could then be proved that

$$\chi(G;x,y) = \sum_T x^{r(T)} y^{s(T)}, \qquad (5.26)$$

where T runs through the spanning trees of G. A proof had to be included of the curious theorem that $\chi(G)$, thus defined, was independent of the particular enumeration imposed on the edges. Details can be found in [74].

This was pleasing, in that it showed the close relationship between chromatic polynomials and the Four Colour Problem on the one hand, and the theory of squared rectangles on the other. Yet I had misgivings. No more results like Smith's Theorem seemed to be arising out of all this algebraic activity. Chromatic polynomials were introduced by G.D. Birkhoff in the hope that they would help settle the Four Colour Problem. But they have not done so. They have not even given an alternative proof of Brooks' Theorem. Students of tetrachromatology meditate upon a hypothetical 5-chromatic map, with a 5-colouring in which only one face has the fifth colour, and they look for ways of removing that unwanted hue. But in the theories of nowhere-zero cycles and of chromatic polynomials they find no suggestion of any such way.

---

[3] A counter-example is now known. It is due to Werner Schwärzler of Bonn. "The coefficients of the Tutte polynomial are not unimodal", *Journal of Combinatorial Theory*, Series B **58**, 240–242 (1993).

When the Haken–Appel proof of the Four Colour Theorem was published one reaction was that the theory of chromatic polynomials had now become pointless. Should it not therefore be abandoned?

But that is not the way of Mathematics. Perhaps at one time the theory of chromials could be justified only as one more device for attacking the Four Colour Problem, but now it goes ahead independently. In compensation for its failure to settle the Four Colour Conjecture it offers us the Unimodal Conjecture for our further bafflement. It may have little to say about the case $m = 4$, but it beguiles us with fascinating tales about the case

$$m = (3 + \sqrt{5})/2.$$

Any survey of algebraic Graph Theory should mention orthogonality. Two 1-chains f and g on a graph G are "orthogonal" if

$$\sum_E f(E)g(E) = 0, \tag{5.27}$$

where the sum is over all edges E of G. It can be shown that the cycles of a graph are those 1-chains that are orthogonal to all the coboundaries, and that the coboundaries are the 1-chains orthogonal to all the cycles. Associated with orthogonality is an algebraic duality in which cycles correspond to coboundaries. It is closely related to the duality of planar graphs.

The graph-functions discussed so far have applied to all finite graphs. But it is usually possible to work in the domain of cubic graphs only. Suppose A is a link of a cubic graph G. Let its ends be x incident with other edges B and C, and y incident with D and E, as in Figure 5.3. We can form a new cubic graph H by "twisting" A, that is by detaching all the edges at x and y and then reconnecting them as shown in the figure. There is a second possible twisting of A that gives the graph K. For planar graphs only one of H and K is normally acceptable, the other destroying planarity.

Figure 5.4 again shows the neighbourhood of A in G, and it shows also the corresponding portion of G'_A. But G'_A is not a cubic graph, for it includes the divalent vertices x and y. However we can make it cubic by "suppressing" x and y. To suppress x we delete that vertex and unite its incident edges B and C to form a single edge BC. The ends of BC must now be those vertices of G, other than x, that are incident in G with B or C. So we suppress x and y, thereby converting G'_A into a cubic graph G_A. The relevant portion of G_A is shown in the third diagram of Figure 5.4.

The preceding definition of G_A cannot be regarded as complete until it has been interpreted for some singular cases. As regards the suppression of x I think there is no difficulty if B and C are links of G with distinct other ends. If those links have the same other end, that is if B and C are

ALGEBRA IN GRAPH THEORY 59

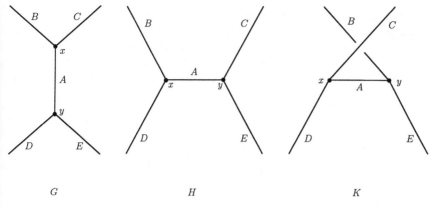

Figure 5.3 Twisting.

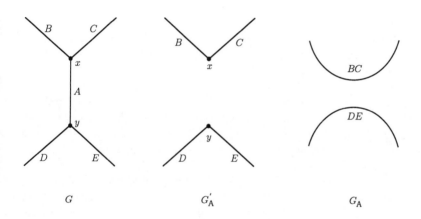

Figure 5.4 Edge-deletion in a cubic graph.

the two edges of a 2-circuit, then BC simply becomes a loop of G_A. There remains the case in which B and C coincide as a loop incident with x in G. Then we say that BC is a "loose edge" of the graph G_A, incident with no vertex at all. We permit such loose edges in our cubic graphs to ensure that edge-deletion shall always be possible. Each loose edge in a cubic graph G is held to constitute, all by itself, a component of G. We even acknowledge "cubic" graphs with no vertices at all, but with one or more edges, all of course loose.

A function f of graphs, with loose edges allowed, is called "topologically invariant" if it is not altered by the suppression of any divalent vertex. The flow-polynomial is one example. It can be shown without difficulty that a graph-function f satisfying conditions (5.11) and (5.12) is topologically invariant if and only if it takes the value -1 for the vertex graph [72].

Let us postulate a topologically invariant graph-function f that satisifies (5.11) and (5.12). Applying it to the graphs of Figures 5.3 and 5.4 we find that

$$f(G) - f(G'_A) = f(G''_A) = f(H''_A) = f(H) - f(H'_A). \qquad (5.28)$$

Hence, by topological invariance,

$$f(G) - f(G_A) = f(H) - f(H_A). \qquad (5.29)$$

This is a recursion formula that involves cubic graphs only.

Suppose we intend to use (5.29) in constructing a catalogue of values of f for connected cubic graphs. We shall find it necessary to know f beforehand for an infinite family of simple special cases. I write the one used in [72] as $\{U_0, U_1, U_2, U_3, \ldots\}$, where the cubic graph U_n is understood to have $2n$ vertices and is further defined as follows. U_0 is the vertexless graph with a single loose edge. In U_n, where n is non-zero, the $2n$ vertices can be enumerated as $v_1, v_2, \ldots, v_{2n}$, so that their incidences with edges can be described as follows. There is a loop L_1 incident with v_1 and a loop L_n incident with v_{2n}. If j is an odd number between 0 and $2n$, then v_j and v_{j+1} are joined by a single link. If j is an even number between 1 and $2n-1$, then v_j and v_{j+1} are joined by exactly two links. The successive rows of Figure 5.5 show U_0, U_1, U_2, U_3 and U_4. The enumeration of vertices, if any, is from one loop to the other. U_0 appears as a small circle, unornamented by vertex-dots.

One of the theorems of [72] tells us that any cubic graph G of $2n$ vertices can be converted into U_n by a sequence of link-twistings. Hence, by (5.29), the difference $f(G) - f(U_n)$ can be expressed in terms of the f-values of some cubic graphs with fewer than $2n$ vertices (if n is non-zero). Accordingly (5.29) can be used to determine $f(G)$ recursively when the values of the $f(U_1)$ are given. In the most general function of cubic graphs satisfying (5.12) and (5.29) the $f(U_1)$ are independent indeterminates.

When I had brought the tree-number, the chromatic polynomial and the flow-polynomial into the theory of "V-functions", that is graph-functions satisfying (5.11) and (5.12), I thought of my work on 1-factors and wondered if they could be brought into the theory too. Let us denote the number of 1-factors of a graph G by $J(G)$. I found that for cubic graphs $J(G)$ did indeed satisfy a simple recursion. It was

$$J(G) + J(G_A) = J(H) + J(H_A) \qquad (5.30)$$

in the notation of Figures 5.3 and 5.4. An equivalent statement is that the function $(-1)^n J(G)$ satisfies (5.29). Both $J(G)$ and $(-1)^n J(G)$ satisfy

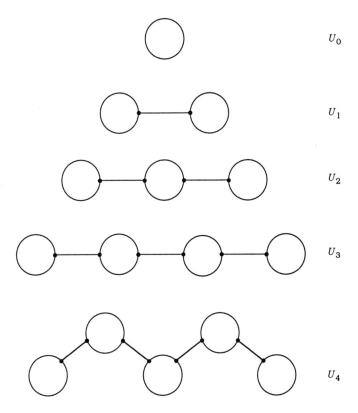

Figure 5.5 The graphs U_i.

(5.12). To make (5.30) completely valid I had to take $J(U_0)$ to be 2. But that is reasonable enough, is it not? U_0 has exactly two subgraphs, itself and the null graph. Each of these is a 1-factor of U_0 in the formal sense that it has the valency 1 at each vertex of that vertexless graph. As for the other U_i it is easy to see that each of them has one and only one 1-factor.

As a trivial example let us use (5.30) to deduce from our values of the $J(U_1)$ the number of 1-factors of the 3-bond B_3, which is a cubic graph. Taking any edge as A we find that $G_A = U_0$, that $H = U_1$ and that H_A has two components each of which is a U_0. So, by (5.30) and (5.12),

$$J(B_3) + J(U_0) = J(U_1) + J(U_0)^2,$$
$$J(B_3) + 2 = 1 + 4,$$
$$J(B_3) = 3.$$

As another check on the recursion formulae let us consider the graphs of Figure 5.6, which shows how the 4-clique can be transformed into U_2 by

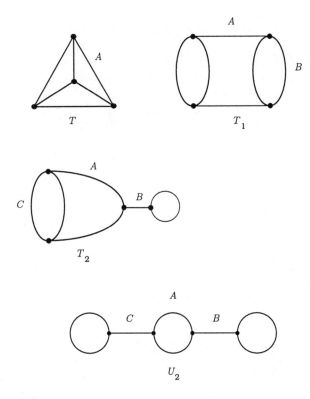

Figure 5.6 Reduction to a U_i by twisting.

three link-twistings. First the link A is twisted to transform the 4-clique T into a graph T_1. Then B is twisted in T_1 to transform it into T_2, and finally C is twisted to change T_2 into U_2. Now $(T_2)_C$ is a U_1, and $(U_2)_C$ has two components, one of which is a U_0 and the other a U_1. Hence

$$J(T_2) = -J(U_1) + J(U_2) + J(U_0)J(U_1) = -1 + 1 + 2 = 2.$$

Next we observe that $(T_1)_B$ is a B_3, and $(T_2)_B$ has two components, one of which is a U_0 and the other a B_3. Hence

$$J(T_1) = -J(B_3) + J(T_2) + J(U_0)J(B_3) = -3 + 2 + 6 = 5.$$

Last, T_A is a B_3 and $(T_1)_A$ is a U_1. Hence

$$J(T) = -J(B_3) + J(T_1) + J(U_1) = -3 + 5 + 1 = 3.$$

All these results are easily verified by inspection.

Another theorem of [72] says that any topologically invariant function of cubic graphs satisfying (5.12) and (5.29) can be extended to all graphs as a topologically invariant V-function. However the extension of $(-1)^n J(G)$, though expressible as a sum over subgraphs, has no obvious relation to the number of 1-factors in the general case. Attempts to find an interesting extension of the function, as distinct from just an extension, were failures. Many years later however I did obtain a generalization that seemed moderately interesting. That work is recorded in [75].

The recursion formula (5.30) for evaluating $J(G)$ is not recommended for planar graphs. Kasteleyn's device for expressing $J(G)$ as a Pfaffian is clearly much more efficient [40].

6

SYMMETRY IN GRAPHS

Chapter 1 tells of how the Team of Four made use of symmetrical subgraphs in their theoretical method for the construction of perfect squares.

They dealt with structures that they called "rotors", graphs with rotational symmetry. A graph qualifies as a rotor if it is connected and has an automorphism θ of order $n > 2$. This implies that the automorphisms $\theta, \theta^2, \theta^3, \ldots, \theta^{n-1}$ are all distinct and that θ^n is the identical automorphism I. Moreover the graph is required to have a set of n "border vertices" $A_1, A_2, \ldots, A_n$, all equivalent under the cyclic group generated by θ. Indeed they are to satisfy

$$\theta A_j = A_{j+1}, \tag{6.1}$$

where A_{n+1} is to be identified with A_1. An example of a rotor of order 4 is shown in Figure 6.1. The operation θ is represented as a rotation of 90 degrees anticlockwise.

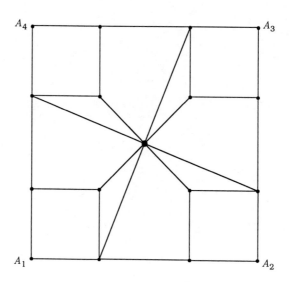

Figure 6.1 A rotor of order 4.

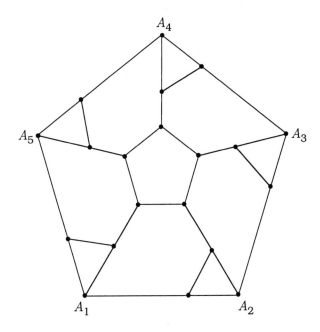

Figure 6.2 A rotor of order 5.

The next Figure, 6.2, shows a rotor of order 5. The rotor of Figure 6.1 has a central vertex, invariant under θ. That of Figure 6.2 has no such invariant vertex.

The graphs shown in these two figures are planar. Non-planar rotors do exist, but they were of no direct interest to the square-squarers. The Four also avoided diagrams with symmetries other than the powers of θ. In particular they avoided symmetries representable as reflections in a line. They wanted a rotor, after the marking of its border vertices, to be distinguishable in the plane from its mirror image. That is the case with Figures 6.1 and 6.2.

Usually the Four were interested in a graph that was the union of two edge-disjoint subgraphs, one being a rotor. The common vertices of the two subgraphs were required to be the border vertices of the rotor. In this chapter I will refer to such a graph as a "rotor–stator combination". The "stator" is the other of the two edge-disjoint subgraphs.

Given a rotor–stator combination we can discuss the operation of replacing the rotor by its mirror image, technically known as "flipping" the rotor. In a planar diagram we may suppose this done by taking the rotor out of the plane, turning it over in three dimensions and then putting it

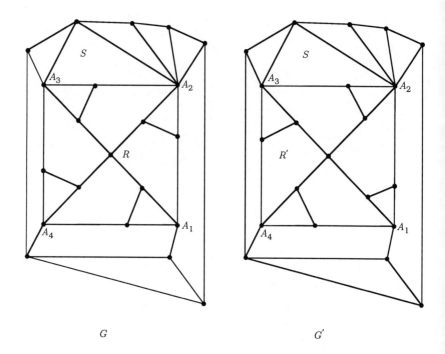

Figure 6.3 Rotor-flipping.

back. The operation is illustrated by the two diagrams of Figure 6.3. We write R for the original rotor, R' for the flipped form and S for the stator. The border vertices are lettered A_1, A_2, A_3 and A_4 as unchanging vertices of the stator. Flipping the rotor transforms the original graph G into a graph G' which, in general, is not isomorphic with G.

Referring back to Figure 1.6 we find essentially the flipping of a rotor of order 3. In this case the stator is made up only of the single edge x and the three border vertices A, B and C. The graphs G and G' happen to be isomorphic. However if we make them into electrical networks by taking A and C as the poles in each case, then they cease to be isomorphic.

The Four noticed that in a rotor–stator combination some of the electrical properties were unchanged when the rotor was flipped. Suppose for example that we take G and G' in Figure 6.3 to be electrical networks of unit resistances, with poles P and Q in the stator S. Then, according to the theory of the Four, all the full currents and full potential differences in the stator persist unaltered when the rotor is flipped. But the currents in the rotor are expected to change. In particular the complexity, the number of spanning trees, of G is invariant under rotor-flipping. It might be interesting to search for other invariants of this operation. At first the Four had no proof of their theory, but thought it justified empirically by the perfect squares to which it led. In time a rather complicated inductive proof was found.

As remarked in Chapter 5 the complexity or tree-number $C(G)$ of a graph G satisfies the recursion

$$C(G) = C(G'_A) + C(G''_A), \tag{6.2}$$

for any link A of G. The same chapter introduces the dichromate $\chi(G; x, y)$ of G as a function satisfying the similar recursion

$$\chi(G; x, y) = \chi(G'_A; x, y) - \chi(G''_A; x, y). \tag{6.3}$$

But now the recursion is valid only when A, an edge of G, is neither a loop nor an isthmus. Many years after the publication of our first paper on squared rectangles I began to wonder what would be the effect of rotor-flipping on the dichromate. If I am to discuss this problem properly I must first generalize the concept of a rotor–stator combination.

We have agreed that each border vertex of the rotor is to be identified with some vertex of the stator, perhaps an isolated vertex of the stator as happened with Figure 1.6. But from now on we allow two or more border vertices of the rotor to be identified with the same vertex of the stator. Thus in place of Figure 6.3 we might have something like Figure 6.4.

Here we use the same rotor and stator as in Figure 6.3. But now one vertex of the stator is used twice, being identified with the two border vertices A_1 and A_2 of the rotor.

Let us now contemplate the case of an edgeless stator each of whose vertices is identified with some border vertex of the rotor. Instead of showing these identifications directly in a diagram it is best to number the vertices of the edgeless stator. Then, in a drawing of the rotor, we mark against each border vertex the number of the stator vertex with which it is identified. An example is shown in Figure 6.5. The rotor is of order 6 and the stator consists of three isolated vertices numbered 1, 2 and 3.

Let us partition the border vertices of a rotor into subsets corresponding to the vertices of the associated edgeless stator. We assume that the stator has no vertex outside the rotor, so that the graphs G and G' are connected.

It is important to observe that if the partition is symmetrical with respect to reflection in some line then the two graphs G and G', related by rotor-flipping, are isomorphic. Figure 6.6 gives an example. Here the partition is symmetrical in the line (1–4). It does interchange vertices 2 and 3 of the stator, but that does not spoil the isomorphism of G and G'. In the two diagrams of this figure the rotor is kept fixed, and the flipping is achieved by reflecting the stator.

Before discussing the effect of rotor-flipping on dichromates let us first be clear as to the recursion formulae for $\chi(G; x, y)$. We have already noted

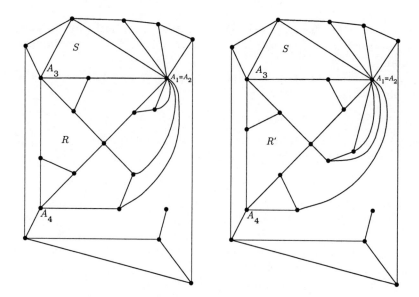

Figure 6.4 Another example of rotor-flipping

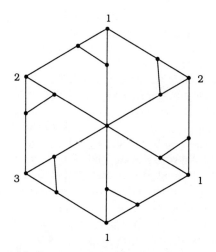

Figure 6.5 Rotor and edgeless stator.

one as equation (6.3). Another appears as equation (5.19) of Chapter 5. It states that if a connected graph G is the union of two edge-disjoint subgraphs H and K with only one vertex in common, then

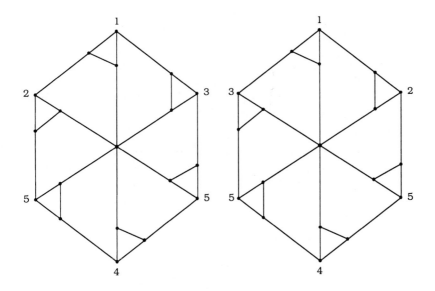

Figure 6.6 An unwanted isomorphism.

$$\chi(G; x, y) = \chi(H; x, y)\chi(K; x, y). \tag{6.4}$$

Note that each of H and K is necessarily connected. We consider only connected graphs G in what follows, and we note that rotor-flipping preserves connection.

As a particular case of (6.4) suppose G is obtained from a graph H by adjoining one loop. Then we can apply (6.4) with K a loop-graph. We find that

$$\chi(G; x, y) = y\chi(H; x, y). \tag{6.5}$$

Next suppose that the connected graph G has an isthmus A. Then G is the union of three mutually edge-disjoint connected subgraphs H, K and L, where L is a link-graph. The graphs H and K are disjoint and each includes one end of L. (See Figure 6.7.) So by (6.4) we have

$$\chi(G) = \chi(H)\chi(K)\chi(L) = x\chi(H)\chi(K),$$

$$\chi(G''_A) = \chi(H)\chi(K), \tag{6.6}$$

$$\chi(G; x, y) = x\chi(G''_A; x, y).$$

Let us now consider a fixed rotor R associated with an arbitrary stator S, subject only to the condition that the rotor–stator combination G is connected. Using the above recursion formulae we proceed to prove the following theorem.

Theorem 6.1 *Either $\chi(G; x, y)$ is unchanged by the flipping of R for every S, or it is changed by the flipping of R for some edgeless S.*

To prove this theorem we assume that $\chi(G; x, y)$ differs from $\chi(G'; x, y)$ for some choice of S. Choose G so that S has the least number m of edges consistent with this condition. If $m = 0$ the theorem is true. We may therefore assume that S has an edge A.

Case I: A is a loop.

In this case the deletion of A from G and G' gives two connected rotor–stator combinations H and H' respectively such that H' is obtained from H by flipping the rotor R. Hence $\chi(H') = \chi(H)$ by the choice of S. But then $\chi(G') = \chi(G)$ by (6.5), and we have a contradiction.

Case II: A is an isthmus.

We partition G into H, K and L as before. Without loss of generality we may suppose the connected graph R to be a subgraph of H. Flipping the rotor replaces H by another graph H', but leaves K and L unchanged. We have $\chi(H) = \chi(H')$ by the choice of G. Hence $\chi(G) = \chi(G')$, by repeated application of (6.4). This contradicts the choice of G.

Case III: A is neither a loop nor an isthmus.

By deleting or contracting A we get the graphs G'_A and G''_A respectively. Each has R as a subgraph, possibly with an extra identification of border vertices in the latter case. Flipping R converts G'_A into $(G')'_A$ and G''_A into $(G')''_A$ respectively. In neither case is the dichromate altered, by the choice of G. Hence we again have the contradiction that $\chi(G) = \chi(G')$, by (6.3).

We conclude that in G the stator must be edgeless. The theorem is proved.

Does the dichromate remain unchanged when a rotor R is flipped with an edgeless stator? It is found that the answer depends on the order m of R. Let us try $m = 3$ as the first case of significance. There are essentially only three different partitions of the set of border vertices, and they are all symmetrical. They are sketched in Figure 6.7. In each diagram the broken line indicates a line of symmetry.

In view of the theorem just proved we conclude that, for a general G, the dichromate $\chi(G; x, y)$ is invariant with respect to the operation of flipping a rotor of order 3.

Figure 6.8 shows the seven distinct partitions of border vertices for a rotor of order 4. They are all line-symmetrical. Flipping a rotor of order 4 in a graph G cannot alter the dichromate of G.

Figure 6.9 shows the twelve partitions for a rotor of order 5. They too are all line-symmetrical.

For rotors of order 6 or more unsymmetrical partitions of the order-set become possible. For example the partition shown in Figure 6.5 for

SYMMETRY IN GRAPHS

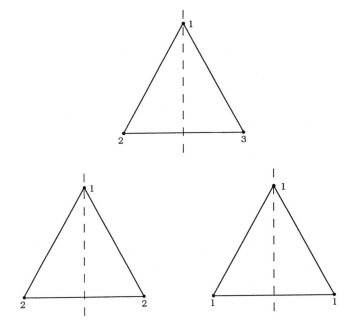

Figure 6.7 Symmetries of edgeless stators (3rd order).

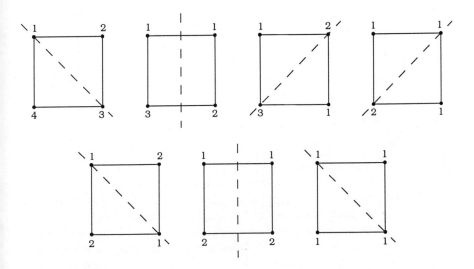

Figure 6.8 Symmetries of edgeless stators (4th order).

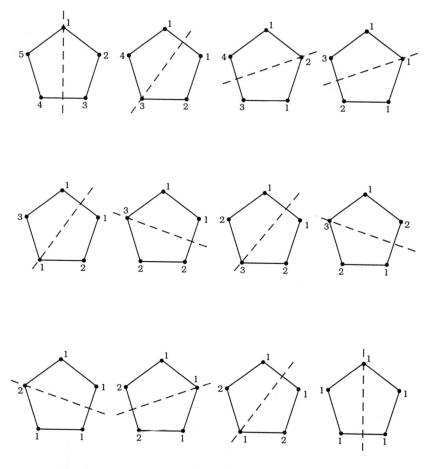

Figure 6.9 Symmetries of edgeless stators (5th order).

a rotor of order 6 is unsymmetrical. We are left with the theorem that the dichromate of a graph is invariant with respect to flipping of rotors of order not exceeding 5. This proof was first published in [80]. S. Foldes has found a non-planar case in which the flipping of a rotor of order 6 changes the chromatic polynomial and therefore the dichromate [26]. But I know of no planar graph in which the flipping of a rotor of any order has been shown to alter the dichromate. The relation between the dichromate and the chromatic polynomial implied here is

$$P(G,\lambda) = \lambda(-1)^{|V(G)|+1}\chi(G; 1-\lambda, 0), \tag{6.7}$$

a result deducible from equations (5.15) and (5.18) of Chapter 5.

Our theorem does reveal the possibility of a pair of non-isomorphic 5-connected planar maps, related by the flipping of a rotor of order 5, that have the same dichromate. F. Bernhart showed that this was possible even among triangulations, that is, maps on the sphere in which all the faces are triangular. His example disproved a conjecture of Ruth Bari that two distinct 5-connected triangulations could not have the same chromatic polynomial. Even the duals of his two triangulations had equal chromatic polynomials, by equation (5.20) of Chapter 5. According to L. Lee even the flipping of a rotor of order 6 cannot alter the chromatic polynomial of a triangulation [43].

We can now assert that the tree-number of a connected graph, being the sum of the coefficients in the dichromate, is not altered by the flipping of any rotor of order not exceeding 5. But the four square-squarers believed that the tree-number is invariant under the flipping of a rotor of any order whatever, and their belief can be justified. They sketch a proof in [11].

Reference has already been made in this book to some quadratic identities involving tree-numbers and full potential differences, identities related to Jacobi's Theorem on the minors of an adjugate matrix. In [11] the symbol $[rs, tu]$ is used to denote the full potential drop in a graph G from vertex r to vertex s when t is the positive pole and u the negative. With a possible adjustment of sign it is the determinant of the submatrix of $K(G)$ obtained by striking out the rows of r and s and the columns of t and u. So $[rs, rs]$ must be equal to the complexity of the graph G_{rs} derived from G by identifying the vertices r and s. A typical quadratic identity goes like this:

$$C(G) \cdot [rs, rs]_{pq} = [rs, rs][pq, pq] - [rs, pq]^2, \tag{6.8}$$

where the suffix pq indicates a function of the graph G_{pq}. Now we can rewrite $[rs, rs]$ as the tree-number of the graph G_{rs}. Likewise $[pq, pq]$ is $C(G_{pq})$. Moreover $[rs, rs]_{pq}$ can be interpreted as the tree-number of the graph $G_{rs,pq}$ obtained from G by identifying first r with s and then p with q. In particular we may have $s = p$, in which case we are discussing the tree-number of the graph G_{rsq} obtained from G by identifying the three vertices r, s and q.

It is also shown in [11] that a "transpedance" $[rs, tu]$ can be expressed in terms of "impedances", that is, transpedances of the form $[\alpha\beta, \alpha\beta]$. The formula is

$$2[rs, tu] = [ru, ru] + [st, st] - [su, su] - [rt, rt]. \tag{6.9}$$

By combining (6.8) with (6.9) we can express the tree-number of a graph obtained from G by two vertex-identifications as a function of the tree-numbers of those graphs obtainable from G by one vertex-identification only. By repeated application of this result we can in theory express the tree-number of a graph got from G by any number of vertex-identifications in terms of those tree-numbers involving only one identification of a pair of vertices.

This theoretical result has an obvious application to rotor-flipping with an edgeless stator. Suppose this flipping to alter the tree-number for any partition of the border vertices. Then some partition involving at most one identification of a pair of vertices must also alter the tree-number. But this is impossible because all partitions of such simplicity are line-symmetrical. We conclude that the tree-number of a graph cannot be altered by rotor-flipping, whatever the order of the rotor. It appears from Foldes' result that chromatic polynomials satisfy no quadratic identities analogous to those for tree-numbers.

Another aspect of graph symmetry came to my attention when I studied Hamilton's "Icosian Game". Hamilton imagined someone tracing a path along the edges of a regular dodecahedron. At each vertex the traveller has the choice of turning right (r) or left (l). If he makes five successive turns to the right (or to the left) he ends up in his starting position. So Hamilton wrote

$$r^5 = I, \qquad l^5 = I. \tag{6.10}$$

Having established a few such identities in r and l Hamilton constructed a (non-commutative) product of 20 r's and l's that was equivalent to the identity, and found that this product was equivalent to what we call a Hamiltonian circuit. All this was described in my recreational textbook as Hamilton's group-theoretical solution of the Hamiltonian circuit problem for the dodecahedron.

It occurred to me to try the method of r and l with some other cubic graphs. But I wanted r and l to be defined in terms of graph structure and not according to orientation on some geometrical or topological surface. In the case of the dodecahedral graph this could be arranged by making the traveller a snake, named "Serpens", exactly two edges in length. Serpens normally lies along two edges, with his head and tail at vertices. Wherever he may be in the graph he notices that he is resting in exactly one pentagon. He defines turning right as advancing one edge-length around this pentagon. Turning left is the other possibility, moving one edge-length out of that pentagon and coming to rest in another. If Serpens, having made one move to the right, thereafter keeps turning left he follows the path shown in Figure 6.10. Serpens observes that $r^5 = I$ and $l^{10} = I$, but l^5 is not the identity. That is not quite according to Hamilton.

SYMMETRY IN GRAPHS

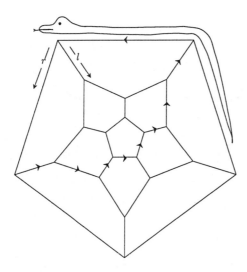

Figure 6.10 Serpens explores his world.

In the same way Serpens can travel around the graphs of the cube and the tetrahedron, finding identities that involve powers and products of r and l.

Things are a little more complicated with the Petersen graph. We find it convenient to adjust its size so that Serpens can lie comfortably along an arc of three edges. He then finds himself to be lying in exactly one pentagon and exactly one hexagon. He defines turning right as advancing one edge-length around the pentagon. Turning left is turning out of the pentagon, and this is found to be advancing one edge-length around the hexagon.

We can use Serpens' adventures in a study of the symmetry of the graph. When he advances one edge-length, either right or left, there is an automorphism that maps his old position onto his new. The diagrams of Figure 6.11 use rotational symmetry to show this clearly, the first for a right turn and the second for a left.

From this observation it follows that all the positions Serpens can reach by a sequence of right and left turns are equivalent under the symmetry of the Petersen graph. It is not difficult to show that Serpens can get to lie along any arc of length 3, in either direction. So all directed 3-arcs in the Petersen graph are alike under the symmetry, a fact already noted without proof in Chapter 3. Like Serpens' initial position, each lies in just one pentagon and just one hexagon. From the uniqueness of the pentagon and hexagon it is possible to show that no non-trivial automorphism of the graph can map any directed 3-arc into itself. It follows that the order of the automorphism group of the graph is the number of its directed 3-arcs,

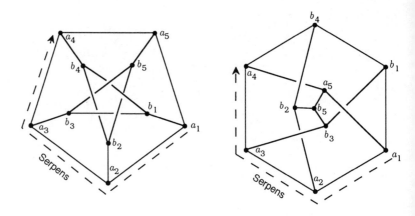

Figure 6.11 Symmetries of the Petersen graph.

that is $10 \times 3 \times 2 \times 2 = 120$. The directed 4-arcs are not all symmetrically equivalent; some of them lie in pentagons and some in hexagons, but none in both.

In my papers, such as [76] and [77], I express the fact that the directed 3-arcs, but not the directed 4-arcs, are symmetrically equivalent by calling the cubic graph "3-regular". Analogously the tetrahedron, cube and dodecahedron have 2-regular graphs.

In searching for other highly symmetrical graphs I was guided by the observations that the tetrahedron is the simplest cubic graph of girth 3 and the Petersen graph the simplest of girth 5. (The girth of a graph is the length of its shortest circuit). The simplest cubic graph of girth 4 is easily found. It must evidently contain the tree of Figure 6.12.

If we can join A, B, C and D by edges so as to transform the tree into a cubic graph without making a short circuit we will have a cubic graph of girth 4 with fewest vertices. This can be done, in essentially only one way. That is by adjoining the circuit $ABDCA$. The resulting graph had been introduced to me as the "Thomsen graph". It is best known as one of the two basic non-planar graphs of Kuratowski's Theorem. It is $K_{3,3}$, the complete bipartite graph of 6 vertices. Two drawings of it appear below in Figure 6.13.

These drawings can be used in a proof that $K_{3,3}$ is 3-regular. In his initial position Serpens lies along three edges, in just one quadrilateral and just one hexagon.

To construct a simplest cubic graph of girth 5 we work similarly with the tree of Figure 6.14. There is essentially only one solution, the Petersen

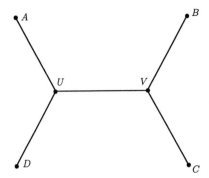

Figure 6.12 Tree for constructing the 4-cage.

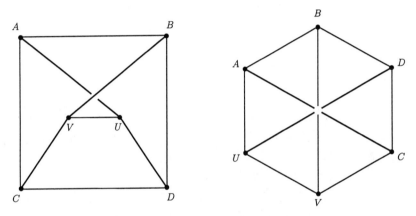

Figure 6.13 Two views of the 4-cage.

graph.

Now let us try girth 6. Evidently the required graph must contain the tree shown in Figure 6.15.

We try to make the graph cubic by introducing joins between the vertices A to H so as to avoid "short" circuits of fewer than 6 edges. Starting with A we have essentially only one choice, the join AE. The apparent alternatives AF, AG and AH would be equivalent to AE under the symmetry of the tree. For a second edge from E we have essentially only the one choice EC, since EB would bring in a quadrilateral. So we go on, constructing the circuit $AECGBFDHA$ in a series of forced moves. The result is the simplest cubic graph of girth 6. It is well known. It is the "Heawood graph", the graph of the seven hexagons on the torus.

78   SYMMETRY IN GRAPHS

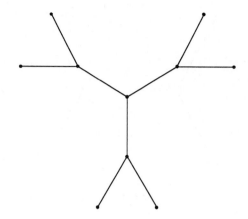

Figure 6.14 Tree for the 5-cage.

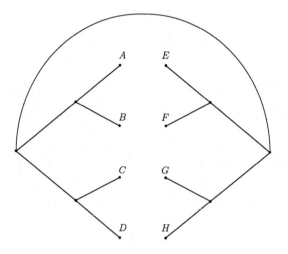

Figure 6.15 Tree for the 6-cage.

By the usual procedure we can verify that the Heawood graph is 4-regular. There are 336 directed 4-arcs, and 336 is the order of the automorphism group.

I tried an analogous construction for girth 7 using the tree of Figure 6.16, which has 22 vertices. Alas, all attempts to join the outer vertices

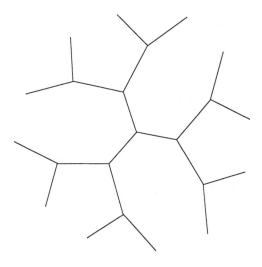

Figure 6.16 Tree for attempting a 7-cage.

so as to make the graph cubic introduced short circuits. Years later W.F. McGee discovered the "7-cage", the simplest cubic graph of girth 7. It has 24 vertices and 32 automorphisms. It is not s-regular for any s. (See [46].)

I went on to girth 8, finding the "8-cage", the graph known to geometers as the Levi graph of the Cremona–Richmond configuration. It has 30 vertices and 1440 automorphisms. It is 5-regular. (See [76] and [79].)

I then had some fun with the algebra of r and l for a general s-regular cubic graph, mainly concerned with the subgroup of symmetries leaving Serpens' tail-tip invariant. I found, somewhat to my surprise, that there would always be a contradiction for $s > 5$. There is no 6-regular cubic graph. (See [76].)

A few years later H.S.M. Coxeter drew my attention to a graph of girth 7 with 28 vertices. It is formed from the three heptagons and seven extra vertices D_i of Figure 6.17. It is completed by joining D_i to A_i, B_i and C_i for each suffix i. The graph is 3-regular. It is also non-Hamiltonian like the Petersen graph. But it does have a Tait colouring. (See [78].)

Some years later I had a phone call from Coxeter. Someone had shown him a symmetrical graph of girth 7. Had I ever heard of it? When he described it I was able to say "Yes, that is what graph-theorists call the Coxeter graph". This incident so aroused or rearoused his interest that he wrote a new paper on the structure, entitled "My graph" ([18].)

One problem arising out of my paper of 1947 remained unsolved for a long time. Is there any 1-regular cubic graph? That is can all the directed 1-arcs be alike but not all the directed 2-arcs? The first example of such a

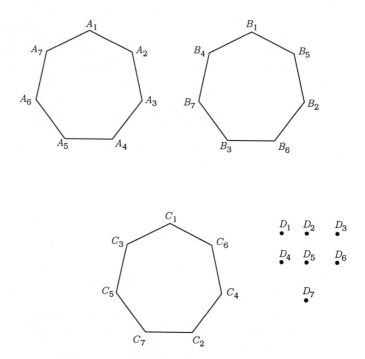

Figure 6.17 The Coxeter graph.

graph was published by R. Frucht in 1952 [27]. Later H.S.M. Coxeter and R.M. Foster agreed that many of the hexagonal maps on the torus have 1-regular graphs.

My last contribution to this theory was in [77], a paper of 1959. I wanted to clear up one minor point left over from 1947, to show that if all the directed 1-arcs of a cubic graph are alike under its symmetry, then the graph is s-regular for some s, and then no non-trivial automorphism can leave a directed $(s+1)$-arc invariant. Later I heard from C. Sims that this extra information had helped him with the theory of permutation groups and that he was making rapid progress with generalizations.

That is where the theory got away from me. But you can find more about it in the writings of C.T. Benson, N.L. Biggs, I.Z. Bouwer, D.Z. Djoković, J. Horton, C. Sims, R.M. Weiss and others.

7

GRAPHS ON SPHERES

Graph Theory, it is said and written, began in 1736. But 3-connected graphs drawn on the sphere are combinatorially equivalent to the convex polyhedra, whose study goes back much further. What are the Platonic solids but highly symmetrical spherical graphs?

Combinatorialists often make a distinction between plane graphs and planar graphs. Plane graphs are figures in the Euclidean plane. Their vertices are distinct geometrical points and their edges are geometrical arcs and circuits, arcs for links and circuits for loops. For planeness no edge must pass through a vertex with which it is not combinatorially incident, and no two edges must meet save at a common incident vertex. A graph is called "planar" if it is isomorphic to a plane graph.

Stereographic projection readily converts figures in the plane into figures on the sphere, and vice versa. So a "planar graph" is a graph isomorphic to one that is realized on the sphere, or as we usually say one that can be drawn on the sphere. So the planar graphs of this chapter are as a rule to be related to the sphere, although our diagrams of them are perforce drawn on a flat surface. A connected planar graph drawn on the sphere partitions the rest of the surface into disjoint simply connected domains called 2-cells, faces or countries. Topologically the vertices, edges and faces are said to make up a 2-complex on the sphere. Combinatorialists usually speak instead of a planar or spherical map.

We are prejudiced in favour of the sphere. After all we are living on a sphere, one that statesmen strive to partition into a stable 2-complex. It is with that 2-complex in mind that we often speak of our graphs on spheres as defining "maps", with "countries" and "borders". The terms "faces" and "edges", derived from the study of convex polyhedra, are at least equally common.

Contemplation of maps led nineteenth-century mathematicians to the Four Colour Problem. They attacked it using the practical knowledge that has to be acquired by all who live on spheres, the knowledge that every circuit has an inside and an outside, and that right can be distinguished from left by reference to markings on the sphere. With this knowledge they reduced the problem to the case of cubic maps having no face with fewer than five incident edges. Using a result of Euler on polyhedra they showed that each of these maps must have at least 12 pentagonal faces. Here we see good progress being made, based on the special properties of 3-connected planar graphs, or equivalently of convex polyhedra.

In the 1880s we find P.G. Tait putting the Four Colour Problem into a new form for cubic maps, working with Tait colourings of the edges instead of 4-colourings of the faces (see Chapter 5). This work seems of interest as offering a way of escape from the sphere into the outer space of general cubic graphs. For Tait's 3-colourings of edges can be discussed for any cubic graph, planar or not.

This historical observation introduces the main theme of the chapter: that some theories belong to the plane, i.e. the sphere, and are developed by specially planar methods; that some of them can be reformulated so as to have meaning for general graphs; and that graphmen should always be on the lookout for such reformulations and be ready to exploit them when they are found. (Or should I say "graphpersons"?)

As an example we can take the purely planar problem of colouring the faces of a planar map in n colours so that no two of the same colour have a common border edge. It is that problem we usually have in mind when we speak of "n-colouring" a map. Dualizing we transform it into the problem of colouring the vertices of a planar graph so that the two ends of any edge have different colours. But that problem can be extended to arbitrary graphs, planar or not. The extension has led to some fascinating conjectures, but to few theorems as yet. One of those few, Brooks' Theorem, deserves mention here.

The form of the theorem which Brooks first showed to the other members of the Trinity Four was concerned with 3-colourings of the faces of a triangulation of the sphere, that is, a map on the sphere in which all the faces are triangles. The tetrahedral graph is a triangulation with four faces, mutually adjacent, and clearly it has no 3-colouring. But it seemed to be exceptional; every other triangulation of the sphere that Brooks investigated was 3-colourable. The form of the theorem that he published in 1941 [10] can be stated as follows.

Brooks' Theorem. *Let k be an integer exceeding 2. Let G be a loopless graph in which no vertex has valency exceeding k, and in which no component is a $(k + 1)$-clique. Then G has a vertex-colouring in k or fewer colours.*

By an "n-clique" we mean a graph of n vertices and $n(n - 1)/2$ edges, each two vertices being joined by a single edge.

I think of Brooks' Theorem as arising, in the form of a conjecture, from a planar assertion by a two-step process. First the theorem is dualized to change it from an assertion about faces and edges into one about vertices and edges. Secondly the restriction of planarity is dropped; the new assertion can be made about any graph, planar or not.

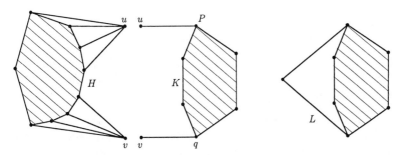

Figure 7.1 Brooks' Theorem (2-separable case).

I now present a proof of Brooks' Theorem by *reductio ad absurdum*. We assume the theorem false. We then suppose a graph G to be chosen satisfying the conditions but not the conclusion of the theorem, and which has the least number of vertices consistent with this condition.

It is easy to establish some results about the connectivity of G. First, G is connected; otherwise each of its components could be k-coloured, by the choice of G, and their k-colourings would combine into one of G. Second, G is 2-connected. For suppose G to be the union of two proper subgraphs H and K, each with at least one edge, that have only a vertex u in common. Then each of H and K has a k-colouring. By permuting colours in K we can arrange that the two colourings agree on u. Then we can combine them to make a k-colouring of G. Third, G is 3-connected. For suppose G is the union of two proper subgraphs H and K, each with at least two edges, having only the two vertices u and v in common. We may hope to colour H and K, to permute the colours in K to make the two colourings agree on u and v, and then to combine the results into a k-colouring of G. Without loss of generality we can say that this procedure will fail only if u and v have the same colour in every k-colouring of H, and different colours in every k-colouring of K. But in that case u and v must have valency at least $k-1$ in H; otherwise we could change the colour of one of them in H. This implies that u and v have valency at most 1 in K, whence by the 2-connection of G they must each have valency exactly 1 in K. We form a graph L from K by identifying u and v. (See Figure 7.1.) Then L satisfies the conditions of the theorem with fewer vertices than G, and therefore L has a k-colouring. This is equivalent to a k-colouring of K in which u and v have the same colour. This contradiction establishes 3-connection.

We next observe that the 3-connected graph G has a vertex-colouring Q in $k+1$ colours. This is because of the valency restriction. The vertices can be coloured one by one, giving each vertex one of the colours to which it is not already joined. We distinguish k of the colours as "ordinary" and number them from 1 to k. The other one, denoted by ω, is "extraordinary" and unwanted. Brooks' device for changing one $(k+1)$-colouring into another is explained in the proofs of the two following lemmas.

Lemma 7.1 *Let u be a vertex coloured ω in Q, and let v be an adjacent vertex. Then there is a $(k+1)$-colouring Q' of G, differing from Q at most on u and v, in which u is not coloured ω.*

To form Q' we first change the colour of v to ω. Then there is at least one ordinary colour to which u is not joined, and we give such a colour to u in place of ω. It may of course happen that v is now joined to another ω. If so we similarly change the colour of v to an ordinary one, one to which it is not joined. We now have the required $(k+1)$-colouring Q'.

Lemma 7.2 *Let L be an arc in G with ends x and y. Then there is a $(k+1)$-colouring Q' of G which differs from Q only on L and in which no vertex of L, with the possible exception of y, has the colour ω.*

To prove this lemma we first enumerate the vertices of L, from x to y, as $u_1, u_2, \ldots, u_p$. We consider these vertices in turn. As soon as we come to a u_i coloured ω we apply Lemma 7.1 with u_i as u and u_{i+1} as v. We then continue along the arc, repeating the procedure until we come to $u_p = y$. We may have to leave y with the unwanted colour. This construction yields the required Q'.

Lemma 7.3 *Let u be any vertex of G. Then we can choose Q so that no vertex other than u has the extraordinary colour.*

To prove this we distinguish, for each vertex x other than u, an arc L_x in G from x to u. We then apply Lemma 7.2 to each arc L_x in turn.

We can use Lemma 7.3 to show that each vertex of G has valency k. For suppose some vertex u to have a lower valency. Then we can choose Q so that no vertex but u has colour ω. We can then change the colour of u to an ordinary one, one to which u is not joined. That contradicts the choice of G.

To complete the proof of the theorem after the manner of Brooks we need two non-adjacent vertices u and v. Such a pair can be found. For otherwise G would be a $(k+1)$-clique, contrary to the choice of G. By Lemma 7.3 we may suppose u to be the only vertex with the unwanted colour.

Now v is joined only to vertices of ordinary colours. Two of these, say x and y, must have the same colour. If x and y, with their incident edges, were deleted the remaining graph would still be connected, by the 3-connection of G. Hence there is an arc L in G from u to v which does not pass through x or y. So by Lemma 7.2 we can transfer the unwanted colour from u to v, arranging that no vertex other than v has this colour and that the colour of x and y persists unchanged. But v is joined to vertices of at most $k-1$ different colours; we can change its colour to an ordinary one to which it is not joined. We thus obtain a k-colouring of G, contrary to the definition of that graph. This contradiction completes the proof.

With Brooks' Theorem still fresh in mind we should look at two related conjectures. First there is Hajós' Conjecture: if a loopless graph cannot be vertex-coloured in k colours it contains a subdivision of a $(k+1)$-clique. This conjecture, once famous in Graph Theory, has been disproved [17]. Then there is Hadwiger's Conjecture: if a loopless graph cannot be vertex-coloured in k colours, then it has a $(k+1)$-clique as a minor. By a "minor" of a graph G is meant a graph derived from G by deleting some edges, contracting others, and then deleting any isolated vertices. Here each "some" and "others" includes "none".

Hadwiger's Conjecture for $k=4$ clearly implies the Four Colour Theorem, since the 5-clique is non-planar. It has even been shown to be equivalent to the Four Colour Theorem [110]. By the Four Colour Theorem we may now say that Hadwiger's Conjecture is proved for $k=4$. For smaller values of k it is trivial.[4]

Another colouring conjecture is concerned with Tait colourings. It asserts that any snark, that is, any cubic graph without a loop or a Tait colouring, contains a subdivision of a Petersen graph. Or equivalently it has a Petersen graph as a minor.

Let us go back to the theory of 0-chains and 1-chains over a ring R that was introduced in Chapter 5. We define the "support" $Sup(f)$ of a 1-chain f as the set of all edges with non-zero coefficients in f. We note that every circuit of a graph G, regarded as a set of edges, is the support of a cycle of G over R. On the other hand the support of any non-zero cycle of G defines a subgraph in which the valency of each vertex is 2 or more, and such a graph must contain a circuit.

Let us now define an "elementary cycle" on G over R as a non-zero cycle with minimal support. This means that its support contains that of no other non-zero cycle. Then by the preceding observations the elementary cycles on G over R are those whose supports define circuits. We have here an equivalence between a graph-theoretical concept and an algebraic one, between the circuit and the elementary cycle. In my thesis [81] I made a search for such correspondences, trying to present as much as I could of Graph Theory in an algebraic form.

There is a principle of algebraic duality saying roughly that anything you can do with cycles you can also do with coboundaries. Thus we can

---

[4]For a recent proof of the case $k=5$ see Robertson et al. "Hadwiger's conjecture for K_6-free graphs". *Combinatorica* **13** (1993), No. 3, 279–361.

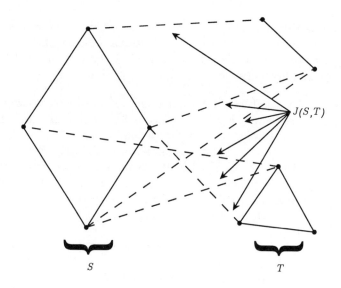

Figure 7.2 The edge-set $J(S,T)$.

define elementary coboundaries analogous to elementary cycles. Let us seek their graph-theoretical interpretation.

Suppose the graph G to be connected. Partition its vertex-set $V(G)$ into complementary non-null subsets S and T. Let $J(S,T)$ denote the set of all edges with one end in S and one in T. (See Figure 7.2.) It is a non-null set, by the connection of G.

We remark that $J(S,T)$ is the support of a coboundary on G over R. For consider the 0-chain f on G over R in which the vertices of S have coefficient 0 and those of T coefficient 1. An edge of $J(S,T)$ has coefficient 1 or -1 in δf, and every other edge of G has coefficient 0.

If we delete from G the edges of $J(S,T)$, the resulting graph is the union of two disjoint subgraphs G_S and G_T. Here G_S is the subgraph of G defined by the vertices of S and the edges having both ends in S, and G_T is defined analogously. If G_S and G_T are both connected we call $J(S,T)$ a "bond" of G. We then call G_S and G_T the "end-graphs" of this bond. If each of the end-graphs is contracted into a single vertex then G is transformed into one of the n-bonds described in Chapter 5, in the discussion of dichromates.

Suppose $J(S,T)$ is a bond of G. It is the support $\text{Sup}(\delta f)$ of the coboundary δf. If δf is not an elementary coboundary there is another 0-chain g such that $\text{Sup}(\delta g)$ is a non-null proper subset of $\text{Sup}(\delta f)$. So some edge A of $J(S,T)$ is not in the support of δg. But all vertices of S have the same coefficient in g, by the connection of G_S and the definition of a coboundary, and an analogous statement holds for G_T. Moreover A has zero coefficient in δg, and therefore the g-coefficients of the vertices of S and T are equal. Hence all the vertices of G have the same coefficient in g and therefore $\delta g = 0$. But this is contrary to the choice of g. We conclude that each bond of G is the support of an elementary coboundary.

There is a converse theorem: the support of every elementary coboundary on G over R is a bond. To prove this consider any elementary coboundary δf, where f is a corresponding 0-chain. Let α be one of the non-zero coefficients occurring in f, let S be the set of all vertices with coefficient α in f, and let T be the set of all other vertices in G. The set $J(S,T)$ is contained in $\mathrm{Sup}(\delta f)$. Since $J(S,T)$ is itself the support of a non-null coboundary δf it follows that

$$J(S,T) = \mathrm{Sup}(\delta f).$$

Now G_S must be connected. For if not let H be one of its components. By the connection of G the set

$$J(V(H), V(G) - V(H))$$

is a non-null proper subset of $J(S,T)$, and like $J(S,T)$ it is the support of a coboundary. This is contrary to the elementary character of δf. Similarly G_T is connected. Hence $J(S,T)$ is a bond.

Combining the two preceding results we see that a set of edges of G is a bond if and only if it is the support of some elementary coboundary on G over R. The theory can be extended to disconnected graphs G if we define the bonds of such a graph as the bonds of its components.

By drawing diagrams one soon convinces oneself that planar duality interchanges bonds and circuits. Thus let G and H be dual planar graphs drawn in the usual way, as in Figure 7.3. If K is a circuit in G the edges crossing it define a bond in H. Each residual domain of the circuit contains an end-graph of the bond.

Planar duality draws attention to algebraic duality in the plane, but algebraic duality applies to all graphs, planar or non-planar. In general it does not produce a dual graph of a given graph G but it does indicate a correspondence between the cycles and the coboundaries of G. Algebraic duality relates nowhere-zero cycles and nowhere-zero coboundaries, much as planar duality relates face-colourings and vertex-colourings.

A Hamiltonian circuit in G is a circuit whose length attains the obvious upper bound, the number of vertices. The corresponding object under algebraic duality is the "Hamiltonian bond". It is a bond $J(S,T)$ in a connected graph G with the greatest number of edges consistent with the connection of G_S and G_T, a bond such that G_S and G_T are both trees.

Grinberg's Theorem about Hamiltonian circuits in a planar graph G readily dualizes to a theorem about Hamiltonian bonds in the dual planar

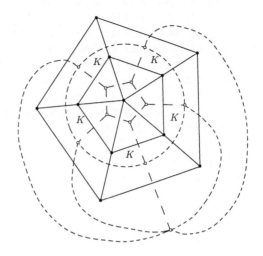

Figure 7.3 Duality of bond and circuit.

graph K. It goes as follows. Let $J(S,T)$ be a Hamiltonian bond of K. Let f'_i be the number of vertices in the tree G_S whose valency in G is i, and let f''_i be analogously defined for the tree G_T. Then

$$\sum_i (f'_i - f''_i)(i-2) = 0. \tag{7.1}$$

This is a reformulation of Grinberg's Theorem. But it makes no implication of planarity. It is meaningful, whether true or false, for any Hamiltonian bond in an arbitrary graph G. Actually it is true in this general case, by essentially the same argument, in dual form, as that used for the planar theorem in Chapter 2.

Jordan's Theorem for the sphere is equivalent, in combinatorial matters, to the statement that the planar dual of a circuit is a bond. A circuit separates the rest of a planar map into two connected sets of faces, just as a bond partitions the rest of a connected graph into two connected end-graphs.

The theory of planar maps makes much use of the "bridges" of a circuit C in an arbitrary connected graph G, as we defined them in Chapter 2. The rest of G falls apart naturally into these bridges. They are connected edge-disjoint subgraphs of G, not contained in C, which can meet one

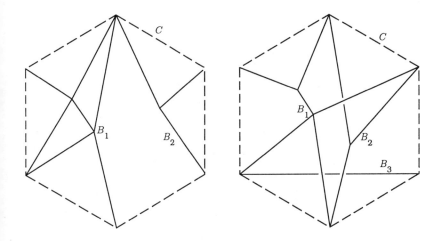

Figure 7.4 Avoiding and overlapping bridges.

another only in vertices of C, and which are minimal with respect to this description. Perhaps their most important property, easily inferred from the definition in Chapter 2, is that two distinct vertices of G can be joined by an arc with no internal vertex in C if and only if they belong to the same bridge. The vertices of a bridge B belonging to the basic circuit C are the "vertices of attachment" of B, and we denote their number by $w(B)$. If $w(B) \geq 2$ the vertices of attachment of B partition C into $w(B)$ arcs, the "residual arcs of B in C". Two such bridges are said to "avoid" one another if all the vertices of attachment of one lie in a single residual arc of the other, a symmetrical relationship. Equivalently they avoid one another if C is the union of two residual arcs, one of each bridge. Otherwise they are said to "overlap". The first diagram of Figure 7.4 shows two bridges B_1 and B_2 that avoid one another. The second shows three mutually overlapping bridges B_1, B_2 and B_3. Here B_3 is degenerate, with no vertex not in C.

These definitions become important when we try to draw G in the plane, or on the sphere. There are two main rules. Any bridge over C has to be drawn within one residual domain of C, except of course that its vertices of attachment must be on C. Second, overlapping bridges must be drawn in different residual domains. From these rules we can get theorems, such as the one saying that three mutually overlapping bridges imply non-planarity. These theorems can be exploited in proofs of Kuratowski's Theorem. (See Chapter XI of [83].)

Let us define a "peripheral circuit" of a graph G as a circuit C having exactly one bridge. There is an interesting theory of peripheral circuits for a 3-connected graph G, which need not be planar. We suppose G to have

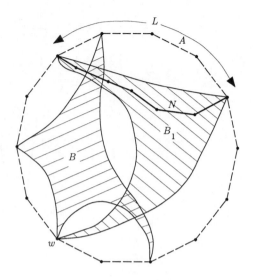

Figure 7.5 Seeking a peripheral circuit.

at least two circuits.

Consider the problem of finding a peripheral circuit through a given edge A. We could begin the search by taking the first circuit C through A that comes to hand and studying its bridges. We may suppose there are at least two of these, for otherwise our search would be at an end. We find moreover that each bridge B over C overlaps at least one other, because of the 3-connection. Otherwise G would fall apart into two subgraphs, each with at least two edges, and with only two vertices in common. These two vertices would be the two ends of a residual arc of B in C, a residual arc bearing the vertices of attachment of at least one other bridge.

We choose a bridge B over C. It overlaps some other bridge B_1. Let L be the residual arc of B_1 containing A, and let L' be its complementary arc in C. (See Figure 7.5.)

Since B overlaps B_1 the vertices of attachment of B include at least one internal vertex w of L'. We can get a circuit C_1 as the union of L and an arc N in B_1 meeting C only at the ends of L. Evidently C_1 has a bridge D containing w, and with it both B and L'. Thus D has more edges than B. If C_1 is not peripheral we repeat the process with C_1 replacing C and D replacing B, and so on. The procedure must terminate with the required peripheral circuit, P say.

There is a curious corollary to this construction. The circuit P has only a single bridge B', and the two ends of A are among the vertices of attachment of this bridge, by 3-connection. It follows that these two vertices are joined by an arc N in B' that has no other vertex in P. Combining this arc with A we obtain a circuit C' meeting P only in A and its two ends. We now repeat the original construction with C' replacing C and with the bridge of C' containing $P - A$ replacing B. It results in another peripheral circuit Q through A that meets P only in A and its two ends. So we can find two distinct peripheral circuits P and Q of G through any given edge A that have only A and its two ends in common.

For a planar 3-connected graph G this result is not surprising. A peripheral circuit must bound a face since the rest of G, being one bridge, must go into a single residual domain. Conversely the bounding circuit of any face must be a peripheral circuit. For if that circuit had two or more bridges one of them would have to overlap another and therefore each residual domain of the circuit would have to contain a bridge. Our result about peripheral circuits thus reduces to the familiar rule that if two faces in a 3-connected map have a common border edge then their bounding circuits meet only in that edge and its two ends. But the result applies to non-planar graphs as well.

We note that for a 3-connected graph G realized in the sphere the boundaries of the faces are determined by the combinatorial structure of G, being the peripheral circuits of that graph. This is Whitney's famous theorem that a 3-connected planar graph can be drawn on the sphere in essentially only one way [116].

We note also that a graph with three distinct peripheral circuits through a given edge A must be non-planar, since an edge of a planar map cannot be incident with three distinct faces. It is even possible to get a converse theorem: in every 3-connected non-planar graph there are three distinct peripheral circuits with a common edge (Theorem XI.59 of [83]).

Planar duality sets up a correspondence between loops and isthmuses, and between contractions and deletions of edges. When a link A is contracted to a single vertex in a planar graph G each circuit through A persists as a circuit in the new graph H, although with one edge fewer. Such a circuit K corresponds to a bond B in the dual graph G^* of G. To get H^* from G^* we have only to delete the edge corresponding to A. The deletion changes B into a bond of H^* with the same end-graphs as before.

Contraction of a loop is a more complex process. A loop, together with its single incident vertex, constitutes a circuit of one edge. Its dual is a bond of one edge, and that edge is an isthmus. Contraction of a loop A to a single vertex v is best thought of as so constricting the sphere that the inside and outside of the circuit A become separate topological spheres, except that they still have the one point v in common. The new graph lies partly one one sphere and partly on the other. The duals of the two parts, each in its own sphere, are the end-graphs of the original one-edge bond.

Any circuit K of G can be contracted to a single vertex v. We contract each edge in turn until the contraction of the last edge, now a loop, fissions the sphere into two spheres. The final graph H can be partitioned into two subgraphs P and Q, one on each sphere and with only the vertex v in common. Their dual graphs are the end-graphs of the bond B of G^* dual to K.

If G is 3-connected it can be verified that the bridges of K are transformed into the blocks of P and Q. We recall that a "block" of a graph is a maximal non-separable subgraph. To this definition we add the convention that a block must have at least one edge. Accordingly we say that the vertex-graph has no blocks. It is well known that planar duality converts the blocks of any planar graph into the blocks of its dual graph. So planar duality converts the bridges of K in G into the blocks of the end-graphs of B in G^*.

These observations suggest a dual concept to that of the bridges of a circuit. We define the "bridges" of a bond B in a graph G, planar or non-planar, as the blocks of its end-graphs. The correspondence under planar duality with the bridges of a circuit is not perfect, except in the 3-connected case, but I have found the definition simple and satisfactory. A "peripheral" bond can be defined as one having a single bridge. Then one end-graph of the bond must be a vertex-graph; the bond consists of all the edges incident with a particular vertex. For 3-connected graphs there is a converse result; we can say that a bond is peripheral if and only if it is such a set of edges.

Figure 7.6 shows a bond B in a graph G, with its two end-graphs H and K. A bridge Q of B is distinguished. It is a block of K, joined to the rest of that end-graph only at the four vertices a, b, c and d. If the edges of Q were deleted from K then these four vertices would be in different components of the resulting graph L, for otherwise the non-separable subgraph Q of K would not be maximal in K. Let us call the components of L containing a, b, c and d the "outgrowths" of Q at a, b, c and d respectively. The outgrowth at any other vertex x would be defined as the corresponding vertex-graph. But the ougrowths at a, b, c and d have edges. Each is a union of blocks of K other than Q, and these blocks are also the blocks of the outgrowth.

Let x be any vertex of Q, and let $J(x)$ be the outgrowth at x. Let $S(x)$ be the set of all edges of B that have an end in $J(x)$. Sometimes $J(x)$ may be null. But if G is non-separable then at least $J(a), J(b), J(c)$ and $J(d)$ must be non-null. In any case the non-null sets of the form $S(x)$ are disjoint and their union is B. They are the members or "parts" of the "partition $\Pi(B, Q)$ of B determined by Q". I think of them as analogous to the residual arcs of a bridge over a circuit. Accordingly I say that two distinct bridges Q and R of B "avoid one another" if there is a part M of $\Pi(B, Q)$ and a part N of $\Pi(B, R)$ such that B is the union of M and N.

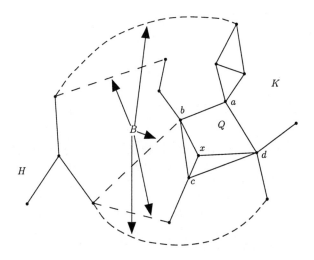

Figure 7.6 Bridges of a bond.

If this is not so I say that Q and R "overlap". It is an interesting exercise in Graph Theory to show that two distinct bridges of B belonging to the same end-graph must avoid one another.

The theory of bridges of bonds is a beautiful theory needing applications. In my papers I have applied it only once. That was to the theory of graphic matroids [81].

8
THE CATS OF CHESHIRE

Looking back on the activities of the Team of Four at Cambridge I note a tendency to put their work into algebraic form. It was Smith who led in this process of algebraicization.

He would tire of his drawings of cubic graphs. He would seek to abolish edges and vertices, preferring to think of a set of points grouped in threes. That may seem a small matter, for the points were but the old edges and the groups of three were the old vertices. But it was the first downward lurch on a slippery slope.

Soon the original edges had become distinguished points in a vector space mod 2. For a while the original vertices persisted as triads of such points summing to zero. But soon they vanished and only linear relations, not necessarily by threes, remained.

Tait colourings took a new and strange form. Instead of a Tait cycle Smith now had an assignment of 1's and 0's to the distinguished points, conforming to the linear relations of those points. A Tait colouring became a set of three of these assignments, summing to zero and putting two numbers on each distinguished point.

What Smith had left he called a "3-net" N. He had a way of defining a 3-net N' whose points were the Tait cycles of N. He would prove such theorems as that the ranks of N' and N'' are equal, that N has a natural embedding in N'', and that N' and N''' are isomorphic. For Smith's account of this theory in its fully developed form see [54].

We others watched with trepidation this process of evaporation and attenuation whereby our cherished visible diagrams went up in a smoke of abstract algebra. We were reminded of a character in English literature, the Cheshire Cat. He was distinguished from other cats by his wide grin and his habit of vanishing into thin air. On one occasion, it is written, the Cat vanished quite slowly, beginning with the tip of his tail and ending with the grin, which remained some time after the rest of him had gone. So when we tried to describe Smith's abstractions to our mathematical friends we would say: "The cat has vanished, but the grin remains."

Through the theory of 3-nets, and through the associated algebraic proof of Smith's Theorem, described in Chapter 5, I too became interested in the algebraic properties of graphs. I was influenced too by a course in Combinatorial Topology. But I took a somewhat different approach to algebraic Graph Theory, concentrating on the theory of cycles and coboundaries over a ring R, the main subject of Chapter 5.

In this theory the vertices of a graph disappear from view. There is a group of 1-chains on the set of edges, a subgroup of cycles, and another subgroup of coboundaries. There are algebraic constructs called elementary cycles and elementary coboundaries, and these are found to correspond respectively to the circuits and bonds of the original graph. It is not difficult to show that the two subgroups are orthogonal. Indeed each consists of all those 1-chains that are orthogonal to all the 1-chains of the other.

I tried to find other graph-theoretical properties that could be characterized algebraically, preferably in terms of cycles and coboundaries. Colourings could be made to depend on nowhere-zero cycles and coboundaries. Deleting an edge A from a graph replaced its cycle-group by the group of those cycles whose supports did not include A. Contracting A made the corresponding replacement for the coboundary group.

I thought it time for the graph-structure to vanish. In place of the edges I kept a set S of objects called "cells". I took the additive group of all chains on S over R, and an arbitrary subgroup N of this, referred to as a "chain-group". Some choices of the ring R were of special interest, for example the ring of integers and the ring of residues mod 2. These gave "integral" and "binary" chain-groups respectively. But usually I made R as general as possible, consistently with commutativity and the existence of a unit element.

Usually I required the chain-group N to be closed under multiplication of its chains by an element of R, although some theorems could be proved without this. I will assume this closure from now on.

It was now possible to erect a theory of chain-groups in analogy with parts of Graph Theory. For example one could define an "elementary chain" of a chain-group N as a non-zero chain of N with minimal support, that is, with a support properly containing the support of no other non-zero chain of N. Such an elementary chain was analogous to a circuit in one part of Graph Theory and to a bond in another. One could then endeavour to extend to chain-groups those results of Graph Theory that were expressible in terms of circuits and bonds.

The notions of subgraphs and contractions have their analogues. Suppose U to be any subset of S. Let us write $N \cdot U$ for the set of restrictions to U of the chains of N. It is a chain-group on U over R. I call it the "reduction of N to U". Next let $N \times U$ be the set of restrictions to U of those chains of N whose supports are contained in U. It is another chain-group on U over R, one that I call the "contraction of N to U".

For an analogy with Graph Theory let us consider a graph G with edge-set S. Let $G \cdot U$ denote the subgraph defined by the edges in U and their incident vertices. Let $G \times U$ denote the graph obtained from G by contracting all those edges that are not in U. Let $\Delta(H)$ and $\Gamma(H)$ be the groups of coboundaries and cycles respectively on a graph H over R. For $H = G$ they are of course chain-groups on S over R. Then, in a simple exercise, one can prove the following rules:

$$\Delta(G \cdot U) = \Delta(G) \cdot U, \tag{8.1}$$

$$\Delta(G \times U) = \Delta(G) \times U, \tag{8.2}$$

$$\Gamma(G \cdot U) = \Gamma(G) \times U, \tag{8.3}$$

$$\Gamma(G \times U) = \Gamma(G) \cdot U. \tag{8.4}$$

These identities justify the assertion that contraction and reductions of chain-groups correspond, more or less, to subgraphs and contractions of graphs.

Given a chain-group N we can partition its cell-set S into disjoint non-null subsets $S_1, S_2, \ldots, S_k$ such that the support of each elementary cycle meets only one of them, and which are minimal with respect to this property. This partition is unique. I call the chain-groups $N \times S_i$ the "components" of N. Each of them is identical with the corresponding $N \cdot S_i$. Moreover each of the $N \times S_i$ has as its elementary chains the restrictions to S_i of those elementary chains of N whose supports meet, and are therefore contained in, S_i. The components of N correspond not to the components of a graph but to its blocks.

Sometimes we can get theorems about chain-groups by generalizing theorems about graphs. For example a theorem of Whitney in [117] implies that any two edges of a block in a graph lie in some common circuit, necessarily contained in that block. This may suggest that any two cells in the same component of a chain-group N are both in the support of some elementary chain. That suggestion can be verified.

Conversely if we have established a general theorem for chain-groups we can derive a theorem about graphs by applying it to the cycle-group of a general graph G. Moreover we can then apply it to the coboundary group of G and get, in general, a different graph theorem. This is why theorems about graphs tend to occur in algebraically dual pairs.

The "dual" of a chain-group N on S over R can be defined as follows. It is the set of all chains on S over R that are orthogonal to all the chains of N. Let us denote it by N^*. We can go on to define the dual N^{**} of N^*. This obviously contains N, but it need not be identical with N. Suppose for example that R is the ring of integers, S has just two cells A and B, and that N consists of the chain $2A + 2B$ and its integral multiples. Then N^* consists of $A - B$ and its integral multiples. Hence N^{**} consists of $A + B$ and its integral multiples. Accordingly $A + B$ belongs to N^{**} but not to N. However in all cases $N^{***} = N^*$.

It is pleasing to have N^{**} identical with N. One way in which this can happen is as follows. Those elements of R that have reciprocals, the unit element for example, are called its "regular" elements. I call a chain of N "primitive" if it is elementary and, except for zeros, has only regular coefficients. It may happen that every elementary chain of N is a multiple of a primitive chain by an element of R. In that case we call N a "primitive" chain-group. It can then be shown that $N^{**} = N$.

If R is a field then all its non-zero elements are regular, and all chain-groups over R are primitive. Moreover all the elementary chains of such a chain-group are primitive. However there are other cases of primitivity.

There are for example primitive chain-groups over the ring I of integers. The regular elements of this ring are 1 and -1, so only these two coefficients can occur in the support of a primitive chain. In my papers I speak of the primitive chain-groups over I as the "regular" chain-groups. It is easily verified for example that the cycle-group over I of a graph G is regular, with primitive chains corresponding to the circuits. Likewise the coboundary group of G over I is regular, with primitive chains corresponding to bonds.

It can be shown that the dual of a regular chain-group is regular. Moreover any contraction or reduction of a regular chain-group is regular (see [84]). The cycle-group and coboundary group of a graph are always duals of one another.

There are two identities involving the duals of reductions and contractions that are important in the calculus of chain-groups. They are as follows:

$$(N \cdot U)^* = N^* \times U, \qquad (8.5)$$

$$(N \times U)^* = N^* \cdot U. \qquad (8.6)$$

In my papers I defined a "dendroid", that is something like a tree, of N as a subset D of S meeting the support of every non-zero chain of N, and minimal with respect to this property. It can be shown that the dendroids of N^* are the complements in S of the dendroids of N. It is found that for a connected graph G the dendroids of $\Delta(G)$ are those sets of edges that define spanning trees of G. The dendroids of $\Gamma(G)$ are of course the complementary sets. If G is not connected a dendroid of $\Delta(G)$ corresponds to a union of spanning trees, one for each component.

There are ways in which one dendroid can be transformed into another, and a theorem saying that all the dendroids of a given chain-group N have the same number $r(N)$ of cells. It seems natural to call $r(N)$ the "rank" of the chain-group. Two useful properties of rank are stated below.

$$r(N \times U) + r(N \cdot (S - U)) = r(N), \qquad (8.7)$$

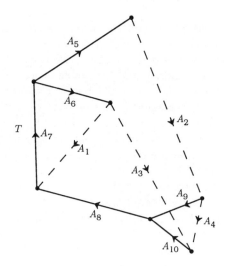

Figure 8.1 A dendroid of a bond-matroid.

$$r(N^*) = |S| - r(N). \tag{8.8}$$

The author apologizes for leaving out so many proofs of theorems on chain-groups. Proofs can be found in [82]. Most of them are assertions or adaptations of standard results in linear algebra.

Consider a dendroid D of a chain-group N on S over R. Let A be a cell of D. Then there has to be a chain f_A of N whose support meets D only in A, by the minimality condition in the definition of a dendroid. If N is primitive we can arrange that f_A has regular coefficients; we can further arrange that $f_A(A) = 1$. In what follows we assume this has been done for each cell of D. It is then clear that each chain f_A is elementary. For if another non-zero chain g of N had its support properly contained in that of f_A, then $g - g(A)f_A$ would be a non-zero chain of N with a support not meeting D.

Figure 8.1 shows a graph G with a distinguished spanning tree T. The edges of T are shown by full lines and the four other edges A_1, A_2, A_3 and A_4 by broken lines. These four edges A_i constitute a dendroid D_T of $\Gamma(G)$. This follows from the facts that T has no circuit but each A_i in D_T can be combined with an arc in T to form a circuit Q_i.

In the general case it is convenient to represent the chains f_A, or rather their coefficient-vectors, as the rows of a matrix M. Then M has $r(N)$ rows and $|S|$ columns. The columns correspond to the cells of N, and it is convenient to make the first $r(N)$ of them correspond to the cells of D. By writing the rows in the right order we can then arrange for the first $r(N)$ columns of N to constitute a unit matrix. We call M, with or without these refinements of ordering, a "representative matrix" of N. With those refinements it is a "standard" representative matrix of N associated with the dendroid D.

As an example we show a standard representative matrix M derived from the graph G of Figure 8.1. It represents $\Gamma(G)$ and is associated with the dendroid D_T. The rows represent elementary cycles corresponding to the circuits Q_1.

$$\begin{array}{c} \\ Q_1 \\ Q_2 \\ Q_3 \\ Q_4 \end{array} \begin{pmatrix} A_1 & A_2 & A_3 & A_4 & A_5 & A_6 & A_7 & A_8 & A_9 & A_{10} \\ 1 & 0 & 0 & 0 & 0 & 1 & 1 & 0 & 0 & 0 \\ 0 & 1 & 0 & 0 & 1 & 0 & 1 & 1 & 1 & 0 \\ 0 & 0 & 1 & 0 & 0 & 1 & 1 & 1 & 0 & 1 \\ 0 & 0 & 0 & 1 & 0 & 0 & 0 & 0 & -1 & 1 \end{pmatrix}$$

There are some well-known theorems about a standard representative matrix $M(D)$ of a regular chain-group N associated with a dendroid D. First any square submatrix W consisting of $r(N)$ columns of $M(D)$ has determinant $1, -1$ or 0. Secondly those submatrices W for which the determinant is 1 or -1 correspond to the dendroids of N. Thirdly $M(D)$ is totally unimodular, that is, the determinant of any square submatrix of $M(D)$, of any order, is $1, -1$ or 0. The third theorem is a simple consequence of the first. The full theory can be found, for example, in [84].

Let us write $M'(D)$ for the transposed matrix of $M(D)$. Then it follows from the above results that the determinant of the product $M(D) \times M'(D)$ is the number of dendroids of N. Taking $M(D)$ to be the standard representative matrix of $\Gamma(G)$ set out above we can use this formula to get the number of spanning trees of G. This number can equally well be derived from a standard representative matrix of $\Delta(G)$ corresponding to the dendroid $S - D$ of that chain-group. In that case the matrix product can be shown to be part of the Kirchhoff matrix, and we are back to the Matrix-Tree Theorem.

While I was setting out the preceding theory for my thesis I kept thinking of a paper of Hassler Whitney that was or might soon be relevant [118]. Whitney dealt with structures that he called "matroids". He gave several equivalent definitions for them. The one in terms of independent sets is favoured nowadays. But perhaps the one in terms of circuits fits in best with what I have been saying about chain-groups. On this definition a matroid is given by a set S of "cells" and a family of non-null subsets of S called the "circuits" of the matroid. The circuits are required to satisfy the following two axioms.

I: No circuit is a subset of another.

II: If C_1 and C_2 are circuits having a common cell a, and if another cell b belongs to C_1 but not to C_2, then there is a circuit C_3 such that b belongs to C_3 and C_3 is contained in $(C_1 \cup C_2) - \{a\}$.

It is easily verified that the supports of the elementary chains of a chain-group N are the circuits of a matroid, the "matroid of N". In particular the circuits of a graph are the circuits of one matroid and the bonds of the graph are the circuits of another. We recognize a further stage of abstraction from Graph Theory. Even the algebraic adornments are now gone and we are down to pure Set Theory.

There is no reason to suppose that every matroid has a corresponding chain-group. Yet much of the theory of chain-groups survives in that of matroids. There are dendroids, ranks and duals, reductions and contractions. To each matroid M there is a unique "dual matroid" M^* with dendroids complementary to those of M, and having M as its own unique dual. The theory of matroids is discussed in [82] and [112].

My main preoccupation after passing from chain-groups to matroids was the question: "When is a matroid graphic?" By this I meant: "When is a given matroid the bond-matroid of a graph?" It seemed to me that this problem could conveniently be divided into three subproblems as follows.

(i) When is a matroid binary? That is, when can it be represented as the matroid of a binary chain-group?

(ii) When is a binary matroid regular? That is, when can it be represented as the matroid of a regular chain-group?

(iii) When is a regular matroid graphic?

That is a natural progressive catechism. We know that all graphic matroids, being matroids of chain-groups of coboundaries, are regular. Moreover it can be shown without much difficulty that all regular matroids are binary (see [82]).

In a sense the problem was simplified for me by the observation that there is little difference between a binary chain-group and its matroid. In a binary chain-group we need hardly distinguish between a chain and its support. So if we look only at its elementary chains we see the circuits of its matroid.

I was able to offer some simple answers to the first subproblem. One of them says that a matroid is binary if and only if every mod 2 sum of circuits is a union of disjoint circuits, or is null. Another necessary and sufficient condition is that each contraction of rank 2 shall include at most three circuits. It remained only to decide under what conditions a binary matroid is first regular and then graphic.

Implicit in the results of Whitney was the statement that not all binary matroids are regular. Consider for example the "Fano matroid" M. It corresponds to the binary chain-group N having the standard representative matrix shown below:
$$\begin{pmatrix} 1 & 0 & 0 & 1 & 0 & 1 & 1 \\ 0 & 1 & 0 & 1 & 1 & 0 & 1 \\ 0 & 0 & 1 & 0 & 1 & 1 & 1 \end{pmatrix}$$
The first three columns correspond to the cells of a dendroid D. The chains of N are represented by the linear combinations of the rows of N. There are just seven non-zero ones, all elementary and all with 4-cell supports.

If M is regular it is the matroid of a regular chain-group N_r, and the supports of the elementary chains must be the same for N_r as for N. This implies that N and N_r have the same dendroids. Accordingly the matrix shown above can be changed into a representative matrix of N_r by replacing each residue 0 mod 2 by the integer 0, and each residue 1 mod 2 by one of the integers 1 and -1. We can arrange for the entries in the first three columns to be 0's and 1's only. We conclude that M must have a standard representative matrix, over the integers, of the following form:
$$\begin{pmatrix} 1 & 0 & 0 & x_{11} & 0 & x_{13} & x_{14} \\ 0 & 1 & 0 & x_{21} & x_{22} & 0 & x_{24} \\ 0 & 0 & 1 & 0 & x_{32} & x_{33} & x_{34} \end{pmatrix}$$
Here the entries are integers and each x_{ij} is either 1 or -1. We recall that a standard representative matrix of a regular chain-group must be totally unimodular. By this theorem the determinant
$$\begin{vmatrix} x_{11} & x_{14} \\ x_{21} & x_{24} \end{vmatrix}$$
must be 1, -1 or 0. But it is clearly even and therefore it is zero. We can therefore write
$$x_{11}x_{21}x_{14}x_{24} = 1, \qquad (8.9)$$
and similarly
$$x_{22}x_{32}x_{24}x_{34} = 1, \qquad (8.10)$$
and
$$x_{13}x_{33}x_{14}x_{34} = 1. \qquad (8.11)$$
We next apply the unimodularity theorem to the determinant
$$\begin{vmatrix} x_{11} & 0 & x_{13} \\ x_{21} & x_{22} & 0 \\ 0 & x_{32} & x_{33} \end{vmatrix}$$
which can be expanded as $x_{11}x_{22}x_{33} + x_{13}x_{21}x_{32}$. It is even and therefore zero, so we can write

$$x_{11}x_{21}x_{22}x_{32}x_{13}x_{23} = -1. \tag{8.12}$$

But now

$$-1 = x_{11}x_{21}x_{22}x_{32}x_{13}x_{23}x_{14}^2 x_{24}^2 x_{34}^2$$
$$= (x_{11}x_{21}x_{14}x_{24})(x_{22}x_{32}x_{24}x_{34})(x_{13}x_{23}x_{14}x_{34})$$
$$= 1 \quad \text{by (8.9), (8.10) and (8.11).}$$

This contradiction shows that the Fano matroid is not regular. The preceding piece of algebra comes from [82].

Since the dual of a regular chain-group is regular it follows that the dual of our chain-group N also has no regular matroid. In the terminology of matroids the dual of the Fano matroid is non-regular. The matroid is of course binary, and it has the following representative matrix:

$$\begin{pmatrix} 1 & 1 & 0 & 1 & 0 & 0 & 0 \\ 0 & 1 & 1 & 0 & 1 & 0 & 0 \\ 1 & 0 & 1 & 0 & 0 & 1 & 0 \\ 1 & 1 & 1 & 0 & 0 & 0 & 1 \end{pmatrix}$$

This can be regarded as a standard representative matrix of N^*, with the variation that it is now the last four columns that represent the basic dendroid. It is easy to check that each row of this matrix is orthogonal to each row of our representative matrix of N, and hence to verify that the two matrices represent dual chain-groups. Striking out the columns of the basic dendroid in both the matrices we obtain residual matrices that are transposes of one another. So the non-regularity of N^* could be established by the same piece of algebra as for N.

My solution to the second subproblem was given in terms of excluded minors. A "minor" of a chain-group N is a derived chain-group of the form $(N \cdot U) \times V$. A useful lemma asserts that this derived chain-group can also be written as $(N \times P) \cdot V$, where $P = S - (U - V)$. Minors of matroids can be defined in much the same way. In the case of the matroid of a chain-group the minors of a matroid correspond just as one would hope to the minors of the chain-group. There is a theorem that the minors of a regular matroid are regular, and of course the minors of a binary matroid are binary. Contemplation of equations (8.1)–(8.4) assures us that the minors of a graphic matroid are graphic. My conclusion was that a binary matroid was regular if and only if no minor of it was a Fano matroid or the dual of a Fano matroid. Equivalently a binary chain-group is regular if and only if no minor is a Fano chain-group or the dual thereof; that is, if and only if no minor has one of the two binary representative matrices discussed above. The theorem is proved in [85] and [86], and in [82].

For the third subproblem I adapted the theory of bridges, described in Chapter 7, to binary matroids and chain-groups. The resolution of the subproblem was analogous to a proof by bridges of Kuratowski's Theorem. It is described in [87] and [82]. I will conclude the present chapter by describing and discussing the extended bridge theory.

Let N be a binary chain-group on S. Let J be a circuit, that is, an elementary chain, of N. It is convenient to identify any chain of N with its support, so we can treat J as a subset of S. Consider the reduction $N \cdot (S - J)$. We resolve this chain-group into its components. These components are by definition the "bridges" of the circuit J in N.

Let B be one of these bridges, and let L be any circuit of B. Then L is the restriction to the cell-set of B of a chain K of N. Let us note that there is no non-zero chain of N in $J \cup L$ other than J, K and $J + K$. For any such chain P would either contain L or be contained in J, since L is elementary in B. Hence either P or $P + K$ would be contained in J, and so would be either zero or identical with J. So P would be either zero, J, K or $J + K$. There are two possibilities. The first one is that L may be a circuit of N, and so be identical with K. Then the third non-zero chain of N, after the circuits J and L, is the non-elementary chain $J + L$. The other is that K and $J + K$ are circuits of N meeting J in complementary non-null subsets U and V. We distinguish the two possibilities as the "disconnected" and "connected" cases respectively.

In my papers I used a geometrical terminology. Circuits were "points". The above set $J \cup L$ was a "line", "disconnected" in the first case and "connected" in the second. In more general Matroid Theory a "k-flat" Q was a union of circuits such that the corresponding contraction had rank k. In the cases $k = 1, 2$ and 3 I would speak of a "point", "line" or "plane" respectively. The proof in [85] and [86] of the Homotopy Theorem depends on auxiliary theorems about the interrelations between connected and disconnected k-flats for varying values of k.

Returning to our binary chain-group N we observe that if the line $J \cup L$ is connected, that is, if L is not a circuit of N, then it partitions J into complementary non-null subsets U and V such that $L \cup U$ and $L \cup V$ are circuits of N. These two circuits, together with J, give us the three points on our connected line. I call U and V the "primary segments" of J determined by L.

We define a "B-segment" of J as a minimal non-null intersection of primary segments of J determined by circuits L of B. So B partitions J into disjoint non-null B-segments.

Now we can define the terms "avoid" and "overlap". Two bridges B_1 and B_2 of J avoid one another if there exists a B_1-segment W_1 and a B_2-segment W_2 such that $J = W_1 \cup W_2$. In the remaining case B_1 and B_2 are said to overlap. Now in the case of a bridge B over a circuit J of a graph, at least in the non-separable case, the B-segments are found to correspond to the residual arcs of B in J. Moreover each of these B-segments is found to be a primary segment, determined by some circuit of B. In the case of a bridge B over a bond J of a graph, a B-segment turns out to be the set of edges of J having ends in the outgrowth from some vertex x of the block, of one of the end-graphs of J, corresponding to B. Considering the bond of G with this outgrowth as one of its end-graphs we see that, once again, the B-segment is a primary segment. (Such a bond always exists in the non-separable case.)

But in the general case of a binary chain-group a B-segment is not necessarily primary. In the Fano chain-group for example it is found that the B-segments have just one cell each, whereas each primary segment has two.

I think that among the most pleasing theorems of this investigation is the one asseverating as follows. If no minor of N is a Fano chain-group then, for any bridge B over any circuit K, the B-segments are primary segments with respect to circuits of B.

We seek conditions for N to be graphic, that is, to be the coboundary group of a graph. We know however that the bridges of a bond in a graph G are the blocks of two disjoint subgraphs, and there is no overlap between two bridges in the same one of these subgraphs. We conclude that in any graphic chain-group the bridges over any circuit K must fall into two complementary sets such that no two members of the same set overlap. We call such a circuit K "bridge-separable", and if every circuit of N is bridge-separable we say that N is "even". If N is not even it is, naturally, "odd".

The next important theorem on bridges in [82], a rather difficult one, asserts that if N is odd it must contain one of the following chain-groups as a minor: the dual of the Fano chain-group or the cycle-group of one of the two Kuratowski graphs. When this result is established it is fairly easy to prove by induction that if N has neither the Fano chain-group nor one of the three just mentioned as a minor, then it is graphic. Since these four special minors are all non-graphic, and since any minor of a graphic chain-group is graphic, we then have a solution of the third subproblem. A regular binary chain-group is graphic if and only if it has no minor which is the cycle-group of one of the two Kuratowski graphs.

It is possible to give an algorithm to determine whether a given binary chain-group N is graphic and if so to exhibit a corresponding graph. One procedure would be first to reduce to the 3-connected case and then to determine the peripheral circuits, those with a single bridge. These peripheral circuits should correspond to vertices, each circuit being the set of all edges incident with a particular vertex. If three such peripheral circuits ever shared a common edge, then N would be non-graphic. But if there was no such obvious failure the peripheral circuits would determine a graph. It would remain only to establish, perhaps through a general theorem, that N was indeed the coboundary group of this particular graph.

One algorithm for graphicality is presented in [1]. My own effort in this direction is explained in [88].

9

RECONSTRUCTION

Archaeologists may try to assemble broken fragments of pottery to find the shape and pattern of an ancient vase. Chemists may infer the structure of an organic molecule from a knowledge of its decomposition products. Graph theorists try to model such problems of reconstruction in their own subject.

Typically we are shown the graphs that can be obtained from some graph G by deleting one vertex and its incident edges. I will call these the "first" vertex-deleted subgraphs. There are "second" ones, derived from G by deleting two vertices, "third" ones deleting three vertices, and so on. In standard Reconstruction Theory we try to discover how much of the structure of G can in general be inferred from the structure of its first vertex-deleted subgraphs.

A few comments should be made in explanation. If G has n vertices we are to be given diagrams of n vertex-deleted subgraphs, one for each vertex. However the diagram for the ith vertex is to give only the abstract structure of the corresponding subgraph G_i, only its isomorphism class. This means that in our diagrams we are given no correlation of vertices. We are not told which vertex in one diagram corresponds to which in another. A more helpful comment is that we can infer the abstract structure of the second vertex-deleted subgraphs from those of the first. We list all the first vertex-deleted subgraphs of each first vertex-deleted subgraph of G, remembering that each one will appear twice in this list.

The Reconstruction Conjecture, also known as Ulam's Conjecture, asserts that the isomorphism class of G is uniquely determined by the isomorphism classes of the first vertex-deleted subgraphs, provided that G has at least three vertices. The reason for this proviso is easy to see. Suppose G to consist of two vertices and n edges joining them. Then each of the two vertex-deleted subgraphs is a vertex-graph. This fact alone tells us nothing about the value of n.

A graph satisfying the Reconstruction Conjecture is said to be "reconstructible". A property of G, such as connection or disconnection, is said to be "reconstructible" if it can be inferred from the isomorphism classes of the first vertex-deleted subgraphs, whether or not the determination of G is carried to completion.

Around 1950 I was doing some work myself on vertex-deleted subgraphs, but some years elapsed before I tried to apply this work to problems of reconstruction. I was interested in chromatic polynomials. Let me remind you that, for a positive integer λ, the chromatic polynomial $P(G, \lambda)$ of a graph G is the number of its λ-colourings. It is identically zero if and only if G has a loop. In all other cases $P(G, \lambda)$ can be expressed as a polynomial in λ whose degree n is the number of vertices of G, and in which the coefficient of the leading term is 1. The constant term in $P(G, \lambda)$ is always zero except, by convention, in the case of the null graph, the graph that has no edges and no vertices. Excluding the looped and null cases we can write

$$P(G, \lambda) = \lambda^n + a_1 \lambda^{n-1} + a_2 \lambda^{n-2} + \cdots + a_{n-1} \lambda. \tag{9.1}$$

In the work just mentioned it seemed to me convenient to use the reciprocal z of λ, and so to replace $P(G, \lambda)$ by the polynomial

$$f(G, z) = z^n P(G, z^{-1}). \tag{9.2}$$

So I rewrote equation (9.1) as

$$f(G, z) = 1 + a_1 z + a_2 z^2 + \cdots + a_{n-1} z^{n-1}. \tag{9.3}$$

I was interested in the zeros of $f(G, z)$. The corresponding values of λ include all those positive integral cases for which G has no λ-colouring. As I explained in an earlier chapter I knew of an additive (or subtractive) recursion formula for $P(G, \lambda)$. But this did not seem to help in the study of the zeros. It occurred to me that a multiplicative recursion formula, if such could be obtained, might be more powerful. Perhaps one such would relate $f(G, z)$ to the corresponding polynomials for the vertex-deleted subgraphs.

I experimented with some simple graphs. For the vertex-graph, the link-graph and the triangle the f-polynomial is 1, $1 - z$ and $(1 - z)(1 - 2z)$ respectively. The first vertex-deleted subgraphs of the triangle are three link-graphs, and its second ones are three vertex-graphs. I tried multiplying together their polynomials in various ways. I found some amusement in a comparison of the products of f-polynomials for vertex-deleted subgraphs of odd and even orders. For those of odd order the product is

$$(1 - z)^3 = 1 - 3z + 3z^2 - z^3. \tag{9.4}$$

For those of even order, including G itself as it zeroth vertex-deleted subgraph, the product is

$$(1 - z)(1 - 2z) \cdot 1^3 = 1 - 3z + 2z^2. \tag{9.5}$$

How odd that these two products should agree in the first two terms!

Let us treat the tetrahedron similarly. Its first deletions are four triangles, its second ones six link-graphs, and its third ones four vertex-graphs. Its f-polynomial is
$$(1-z)(1-2z)(1-3z).$$
So the even product in this case is
$$(1-z)^7(1-2z)(1-3z) = 1 - 12z + 62z^2 - 182z^3 + \cdots. \tag{9.6}$$
The odd product is
$$(1-z)^4(1-2z)^4 = 1 - 12z + 62z^2 - 180z^3 + \cdots. \tag{9.7}$$

These products agree in the first three terms. To me this was not just amusing. It hinted at a multiplicative theorem. I tested some other reasonably simple graphs and found the same kind of coincidence every time. For any graph G let us write $\Pi_0(G)$ and $\Pi_1(G)$ for the even and odd products, excluding $f(G, z)$ as a factor in the even case. In the cases tested I found that $f(G, z)\Pi_0(G)$ and $\Pi_1(G)$ agreed in the first $n-1$ terms, where n was the number of vertices. I expressed this agreement by writing
$$f(G, z)\Pi_0(G) \equiv \Pi_1(G) \mod z^{n-1}. \tag{9.8}$$
I could even write
$$f(G, z) \equiv \Pi_1(G)/\Pi_0(G) \mod z^{n-1}. \tag{9.9}$$

This is almost a recursion formula for the f polynomial. It says that if we know $f(H, z)$ for each proper vertex-deleted subgraph H of G, then we can calculate the first $n-1$ coefficients in $f(G, z)$. That leaves one unknown coefficient, a_{n-1}, to be determined in some other way.

Excluding trivialities we can assume that G is non-null, loopless and connected. Merely because it has a link we know that it has no 1-colouring. That is, $P(G, 1) = 0$ and therefore $f(G, 1) = 0$. Using this result we can infer the nth coefficient when the first $n-1$ are given. Effectively (9.9) is indeed a recursion formula for chromatic polynomials.

I tried to prove (9.9). I tried to publish the formula, but without success. For one thing, my proof was found to be not fully rigorous. For another, referees thought the result too complicated to be interesting. Who indeed would want to use such massive products in the calculation of a new chromatic polynomial? For myself, I clung to the view that a formula could have theoretical interest without being practically useful, and I thought that this formula did have theoretical interest and was theoretically simple.

From time to time I remembered the result. Eventually I found a proof that I thought rigorous, and which even generalized to dichromatic polynomials. That proof I got published [89] But alas, those for whom it was intended said that they could not follow it. However there it is, as far as I know the first proof that dichromatic polynomials are reconstructible. Moreover I was able to improve upon the analogue of (9.9) by getting a formula for a_{n-1}. This involved, among other things, the number of Hamiltonian circuits of G. It allowed me to assert, some years later, that the number of Hamiltonian circuits is reconstructible.

Soon after this publication I learned that the multiplicative formula for chromatic polynomials was known to some workers in Statistical Mechanics, who even thought it might be useful in calculations. But for physical applications of graph-theoretical polynomials I must refer you to the works of Norman Biggs, and in particular to his Cambridge Tract *Algebraic graph theory* [6].

I began to take an interest in lectures on Reconstruction, still without relating them to my multiplicative recursions. I learned of Kelly's Lemma. Let $s(G, H)$ be the number of subgraphs of G isomorphic with H. The lemma says that if $|V(H)| < |V(G)|$ then $s(G, H)$ is reconstructible. This is because there are $|V(G)|$ first vertex-deleted subgraphs of G, and a given copy of H appears in exactly $|V(G)| - |V(H)|$ of them. To get $s(G, H)$, therefore, we count the number of copies of H in all the first vertex-deleted subgraphs, and divide the total by $|V(G)| - |V(H)|$. Using Kelly's Lemma we can now prove reconstructibility for such things as the number of loops, the number of links and the number of circuits of any length $k < |V(G)|$. From the results about loops and links we can infer the valency in G of the deleted vertex for each first vertex-deleted subgraph. If G is found to have no non-Hamiltonian circuit, then it is itself a circuit if it has $|V(G)|$ edges, and a tree if it has $|V(G)| - 1$. In any case we can determine if G is a regular graph. There are theorems saying that regular graphs and trees are reconstructible.

At one lecture on reconstruction I had a shock. The speaker remarked that there were four important properties of graphs that had not yet been proved reconstructible. They were the chromatic number, the edge-chromatic number, the spectrum and the genus. But I realized that the multiplicative formula would reconstruct the chromatic polynomial, and the chromatic number is merely the least positive integer for which that polynomial is non-zero. Back at Waterloo I discussed the matter with Adrian Bondy and he told me that some progress had been made with the problem of the spectrum. It had been reduced to the problem of reconstructing the number of Hamiltonian circuits. "But that", I protested, "is done in my paper on dichromatic polynomials."

I then wrote a paper "All the King's horses" [90] giving a new version, in a new generalization, of multiplicative recursion. It demonstrates the reconstructivity of, among other things, the chromatic number and the characteristic polynomial, the polynomial whose zeros constitute the spectrum. The title is taken from a book of Lewis Carroll. It is written there that after the fall of Humpty Dumpty "all the King's horses and all the King's men" converged upon the scene, presumably with the intention of putting him together again.

It is not really necessary to use multiplicative recursion to prove dichromatic reconstructibility. This is pointed out by W.L. Kocay in [42]. All can be reduced to Kelly's Lemma. In Chapter 6 I gave the following formula for the dichromatic polynomial of a graph:

$$Q(G;t,z) = \sum t^{p_0(G:S)} z^{p_1(G:S)}, \qquad (9.10)$$

where the sum is over all sets S of edges of G.

Let $N(r,s)$ be the number of spanning subgraphs $G:S$ of G having r components and s edges. Then, using equation (5.13) of Chapter 5, we can rewrite (9.10) as follows:

$$Q(G;t,z) = \sum N(r,s) t^r z^{s-n+r}, \qquad (9.11)$$

the sum being over all relevant values of r and s. So the problem of reconstructing $Q(G;t,z)$ reduces to that of reconstructing the coefficients $N(r,s)$.

At first sight Kelly's Lemma seems not applicable. It counts subgraphs H, in a given isomorphism class, with fewer than n vertices, and we want to count subgraphs with n vertices. But there is a trick whereby Kelly's Lemma can be used to find the number of disconnected spanning subgraphs of G with a specified number of components in each isomorphism class. I explain this trick only for one special case, commenting that the extension to the general case is straightforward.

Suppose then that we require the number of 2-component spanning subgraphs of G whose components are copies of the non-isomorphic connected graphs H and K, which together have n vertices. The number of ways of choosing a copy of H and a copy of K out of the subgraphs of G is reconstructible, being $s(G,H)s(G,K)$, since H and K must each have fewer than n vertices.

If our copies of H and K are disjoint their union is a spanning subgraph of the kind required. In the remaining case their union is a subgraph of G with fewer than n vertices.

In principle Kelly's Lemma tells us how many subgraphs of G, with fewer than n vertices, there are in each isomorphism class. But for a graph L of any of these classes we can discover in how many ways it can be represented as a union of a copy of H and a copy of K. Let this number be $A(L;H,K)$. Summing the product $s(G,L)A(L;H,K)$ over all the possible isomorphism classes of L we obtain the number of ways in which we can choose a copy of H and a copy of K so that these two copies are not disjoint. Subtracting this from $s(G,H)s(G,K)$ we obtain the number required.

Generalizing this argument we find it possible to reconstruct the number of disconnected spanning subgraphs of G of each isomorphism class. We can therefore determine $N(r,s)$ whenever r is 2 or more. But now $N(1,s)$ can be inferred from the obvious equation

$$\sum_r N(r,s) = \binom{|S|}{s}. \tag{9.12}$$

We deduce that the dichromatic polynomial is reconstructible. Hence its specializations, the chromatic polynomial and the flow polynomial, are reconstructible. So therefore is the chromatic number.

Trickery with Kelly's Lemma can be carried further. Suppose for example we want the number of spanning subgraphs of a loopless graph G that have just two blocks, these blocks being copies of two non-isomorphic subgraphs H and K which together have just $n+1$ vertices. We proceed much as before. Since G is loopless H and K must have each fewer than n vertices. There are $s(G,H)s(G,K)$ ways of choosing a copy of H and a copy of K from among the subgraphs of G, and this number is reconstructible. If the copies have only one vertex in common their union is a spanning subgraph of the kind required. In the remaining case their union L has fewer than n vertices. For each possible L-structure we can find the number of occurrences in G by Kelly's Lemma, and by inspection we can find the number of ways of representing it as a union of an H and a K. Summing over all L-structures, each with the appropriate multiplicity $s(G,L)$, and subtracting the result from the product $s(G,H)s(G,K)$ we obtain the number of spanning subgraphs of the kind required.

We can generalize the argument, proving reconstructibility for the number of connected spanning subgraphs with blocks, at least two in number, belonging to specified isomorphism classes.

The number $N(1,s)$ of connected spanning subgraphs of G with s edges can be assumed known, by the preceding argument. The number of them with more than one block is reconstructible, by our generalization. Hence, by subtraction, the number of 2-connected spanning subgraphs of G with s edges is reconstructible. In the case $s=n$ those subgraphs are the Hamiltonian circuits of G.

There is a theory of the decomposition of a 2-connected graph G into "3-blocks". These are k-circuits, k-bonds and 3-connected graphs. Using arguments analogous to those for components and blocks we can discuss the reconstructibility of the number of spanning subgraphs of G having a given decomposition into 3-blocks. This aspect of Reconstruction Theory has been investigated by S. Anacker.

I turn now to the characteristic polynomial of a graph G. The structure of G can be described by an "adjacency matrix" $A(G)$. This is a square matrix of order $n = |V(G)|$. If the vertices of G are numbered from 1 to n we define A_{ij}, the entry in the ith row and jth column of $A(G)$, as the number of edges joining the ith vertex to the jth. Thus a diagonal entry a_{ii} would be the number of loops on the ith vertex. For the loopless graphs usually discussed the diagonal elements are zero. The adjacency matrix of G differs from the negative of the Kirchhoff matrix only in the diagonal elements. The "characteristic polynomial of G" is a polynomial in an indeterminate λ. It is the determinant of the matrix

$$A(G) + \lambda I,$$

where I is the unit matrix. The zeros of the characteristic polynomial constitute the "spectrum" of G. We note that the characteristic polynomial is independent of the particular enumeration of the vertices used in constructing $A(G)$.

The coefficient of λ^j in the characteristic polynomial is the sum of the determinants of the symmetrically placed square submatrices of $A(G)$ of order $n - j$. When $j > 0$ these submatrices correspond to the jth vertex-deleted subgraphs of G, and the sum is therefore reconstructible. To complete the reconstruction of the characteristic polynomial we now need only the constant term, that is $\det(A(G))$.

When we expand the determinant of $A(G)$ we find that its terms correspond to those spanning subgraphs of G in which each component is either a circuit or a link-graph. The number of such subgraphs of each disconnected kind can, as we have seen, be deduced from Kelly's Lemma, and the corresponding terms of the determinant can be evaluated. This leaves only the connected kind, that is the Hamiltonian circuits, since $n > 2$. We have seen that the number of Hamiltonian circuits is reconstructible, and it follows that the characteristic polynomial is reconstructible.

It is sometimes suggested that there may be a graph-polynomial, a function of the isomorphism class of G, whose value uniquely determines that isomorphism class. Various candidate polynomials have been proposed and shot down. Certainly the reconstructibility of such a polynomial would imply the truth of the Reconstruction Conjecture. But those polynomials known to be reconstructible have been proved so by Kelly's Lemma alone. Since the Lemma has not yet proved the Conjecture it is not to be expected that these polynomials will.

There is more than one variety of Reconstruction Conjecture. We have discussed "vertex-reconstruction". There is also "edge-reconstruction". In this we are told the isomorphism classes of the first "edge-deleted" subgraphs of G, that is, the subgraphs that are derived from G by deleting one edge. We are asked to show that in every case these isomorphism classes determine that of G uniquely. The new form of Kelly's Lemma asserts that we can find $s(G, H)$ whenever H has fewer edges than G. It seems to give us a vast amount of information about the structure of G. But it has not yet led to a solution of the Edge-Reconstruction Conjecture.

I have not worked on the Edge-Reconstruction Conjecture myself, and therefore I ought not to enlarge upon it in this book. Nor will I comment on reconstruction problems for digraphs and matroids, beyond remarking that some simple analogues of the graphic Reconstruction Conjecture have proved to be false. (See [15] and [57]).

In "All the King's horses" I was much concerned with the fact that G can be vertex-reconstructed if its characteristic polynomial is found to be prime. In this case the reconstruction can be done with Jacobi's Theorem on the minors of an adjugate matrix. The calculations are analogous to some I did long ago in the theory of squared rectangles. So some kinds of graph can be recognized and then reconstructed. But we knew that already in the cases of regular graphs and trees.

When I survey the present state of Reconstruction Theory I come to the gloomy conclusion that there is nothing to it but exercises on Kelly's Lemma, that something more needs to be established about relations between the isomorphism classes of a graph and its subgraphs, and that this "something more" is likely to be very deep and difficult. I have sometimes tried to prove conjectures relating the isomorphisms and isomorphism classes of graphs and subgraphs, and have always failed. I conclude with a statement of one of these conjectures, one that I have put forward at many Problem Sessions but always without feedback.

This conjecture arose when the Four were working on squared rectangles. Let N and N^* be dual c-nets of a squared rectangle. Let A and A^* be the polar edges of N and N^* respectively. Deleting the polar edges we change N and N^* into the "polar nets" P and P^* respectively. I wanted P and P^* to be isomorphic, for then the squared rectangle would be a squared square. But in all the cases I knew in which this happened the isomorphism extended to one between N and N^*, and as a consequence the squared square was diagonally symmetric and so not perfect. I tried to construct an example in which the isomorphism between the polar nets did not extend to one between the c-nets, but without success. I then stated the conjecture that any isomorphism between the polar nets must necessarily extend to one between the c-nets, wherein of course the polar edges would correspond. It is a plausible conjecture, just as the Reconstruction Conjecture is plausible, and it too is still unsettled.

10

PLANAR ENUMERATION

From time to time in a graph-theoretical career one's thoughts turn to the Four Colour Problem. It occurred to me once that it might be possible to get results of interest in the theory of map-colourings without actually solving the Problem. For example it might be possible to find the average number of 4-colourings, on vertices, for planar triangulations of a given size.

One would determine the number of triangulations of $2n$ faces, and then the number of 4-coloured triangulations of $2n$ faces. Then one would divide the second number by the first to get the required average. I gathered that this sort of retreat from a difficult problem to a related average was not unknown in other branches of Mathematics, and that it was particularly common in Number Theory.

Before I could start work on such a project there was one decision to be made: what sort of triangulation should be studied? There were pathological ones, having loops and squashed triangles, but these were of no interest in map-colouring theory. Granting that triangulations must be loopless we must decide whether or not to permit them to have separating digons. These are 2-circuits with their insides and outsides both triangulated. When separating digons are permitted I speak of "general" triangulations. When they are excluded we have "strict" or "3-connected" ones. Specializing further we come to the "simple" triangulations. These are strict triangulations in which there is no separating triangle, no triangle whose inside and outside are both further triangulated.

From the point of view of a student of the Four Colour Problem the simple triangulations are the most interesting ones, failing some further specialization. But perhaps the kinds with fewer restrictions would be easier to enumerate? Rightly or wrongly I decided to start with the strict triangulations. Figure 10.1 shows the strict triangulations with 2 to 8 faces, counting the outside one.

These are distinct triangulations of the plane, although they would not all be different if drawn on the sphere. Only the first, second and last of these seven strict triangulations are simple.

Having made no progress with the enumeration of these diagrams I bethought myself of Cayley's work on the enumeration of trees. His first successes had been with the rooted trees, in which one vertex is distin-

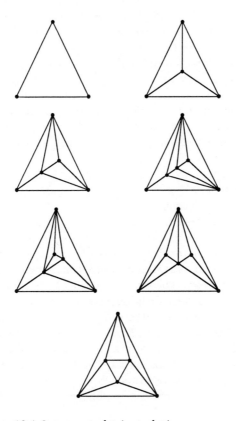

Figure 10.1 3-connected triangulations.

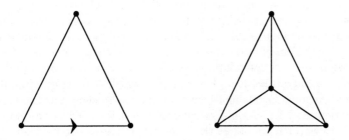

Figure 10.2 Rooted 3-connected triangulations (2 or 4 faces).

guished as the "root". Perhaps I should root the strict triangulations in some way and try to enumerate the rooted ones. Eventually I decided that the rooting should consist of the choice of a face, edge and vertex, mutually incident. Usually I took the root-face to be the outside face and drew an arrow along the root-edge, directing that arrow from the root-vertex. When I came to study more general maps, not necessarily any kind of triangulation, I found it convenient to show a rooting by choosing a root-edge and marking it with two arrows. One was drawn along the root-edge and away from the root-vertex; the other was drawn across the root-edge out of the root-face. This method of rooting could be used even when the root-edge was a loop or an isthmus. In the following diagrams of triangulations the root-face is always the outside one.

Let a_n denote the number of combinatorially distinct rooted triangulations of $2n$ faces. Then we have $a_1 = 1$ and $a_2 = 1$. The corresponding rooted triangulations are shown in Figure 10.2.

We find next that $a_3 = 3$. The three rooted triangulations of 6 faces are shown in Figure 10.3.

Figure 10.4 shows the 13 rooted triangulations of 8 faces. Here a triangulation may be shown with two or more root-arrows. Then we are to choose one of these arrows, and each choice specifies a distinct rooted map. So $a_4 = 13$. I carried the process a little further, finding $a_5 = 68$ and $a_6 = 399$. I am not sure what I would have replied at this stage if I had been asked why I preferred these rooted triangulations to the unrooted ones of Figure 10.1. Later I realized that the distinguishing feature and the great advantage of rooted maps is that they have no symmetry. Automorphisms seem to complicate enumerative problems.

It was natural to write the numbers a_j as the coefficients in a generating function:

$$g(x) = 1 + x + 3x^2 + 13x^3 + 68x^4 + \cdots. \qquad (10.1)$$

Presumably the next step was to find by graph-theoretical arguments an equation satisfied by g. But that seemed to be difficult. Obvious operations such as deleting the root-edge and conventionally rooting what was left usually produced figures that were no longer triangulations. In order to get a class of rooted maps that was closed under such an operation I had to generalize from the triangulations to the "near-triangulations".

A near-triangulation is a planar map in which at most one face is non-triangular. If some face is non-triangular it is to be chosen as the root-face in every permissible rooting. The rooting is best specified by the two-arrow method. However if we adopt the convention that the root-face is always to be the outer face it is usually unnecessary to show the cross-arrow.

I carried over the requirement of strictness to near-triangulations, and I required the outer face to be bounded by a circuit. I also made the rule that

Figure 10.3 Rooted 3-connected triangulations (6 faces).

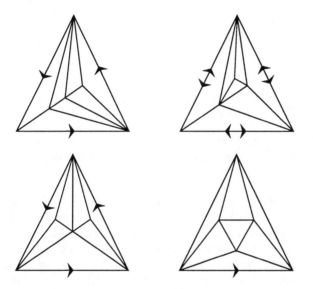

Figure 10.4 Rooted 3-connected triangulations (8 faces).

no internal edge was to have both its ends on this circuit. Perhaps this was a mistake. R.C. Mullin found later that a simpler and equally satisfactory theory could be got without that rule. However the effect of deleting the root-edge is now shown in Figure 10.5.

In the near-triangulation N of this figure the root-edge VW is incident with an internal face VWX. Deleting the root-edge fuses this face with the old root-face and leaves a simple closed curve through X whose interior R is subdivided into triangles. If no edge within this region joins X to another vertex on the new boundary, then the new diagram satisfies the definition of a strict near-triangulation. We root it by taking the outer face as root-face, VX as the root-edge and V as the root-vertex.

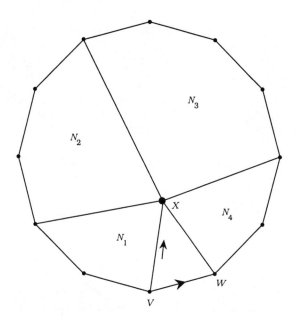

Figure 10.5 Decomposition of a rooted 3-connected near-triangulation.

In the remaining case there are edges $E_1, E_2, \ldots, E_k$ from X across the interior to other vertices on the bounding circuit, as shown in Figure 10.5. They partition the new map into $k+1$ near-triangulations $N_1, N_2, \ldots, N_{k+1}$, which we can root in accordance with some simple convention. The upshot is that N can be constructed from a smaller rooted near-triangulation, or from a combination of smaller near-triangulations, by adjoining a new root-edge so as to cut off a new inner triangle from the root-face. This construction makes it possible to get a recursion formula for the number of rooted near-triangulations with certain specifications, and one finds it possible to express this recursion formula as a functional equation between generating functions. Details are given in [91].

In this early work I defined two numbers associated with a near-triangulation. One was m, defined as three less than the valency of the root-face. The other was n, the number of vertices inside the bounding circuit. The numbers m and n are the "certain specifications" mentioned above. In [91] the number of rooted near-triangulations with given values of m and n is denoted by $\Psi_{n,m}$, and the basic generating function is

$$\Psi = \Psi(x,y) = \sum \Psi_{n,m} x^n y^m, \qquad (10.2)$$

where the sum is over all relevant values of n and m. There is an auxiliary generating function
$$g = g(x) = \sum_n \Psi_{n,0} x^n. \tag{10.3}$$
This is the function already occurring in equation (10.1), the one that enumerates rooted strict triangulations. The functional equation between the generating functions is obtained in [91]. It is
$$(y\Psi + xg - y)(1 - y\Psi) = x\Psi. \tag{10.4}$$

I now know how to solve such equations. But when I first obtained it, I think in 1960, I found it baffling. It was like a quadratic for Ψ, but it was complicated by the presence in the coefficients of the specialization g of Ψ.

Somehow I was able to extract from (10.4) a complicated expression involving one generating function only, namely g. It is discussed in Section 6 of [91]. I used it then to calculate further coefficients in the series g, but I have never used it since. With those new coefficients I could now write

$$\begin{aligned}g = {}& 1 + x + 3x^2 + 13x^3 + 68x^4 + 399x^5 \\ &+ 2530x^6 + 16965x^7 + 118668x^8 \\ &+ 857956x^9 + 6369838x^{10} + \cdots.\end{aligned} \tag{10.5}$$

These coefficients have small prime factors. I feel now that it should not have been difficult to induce from them the general formula

$$g = 2\sum_{n=0}^{\infty} \frac{(4n+1)!x^n}{(n+1)!(3n+2)!}. \tag{10.6}$$

But for reasons I forget I first calculated some coefficients in the inverse function of xg, induced the general coefficient in that function, and worked from there. This happened at the University of Toronto, and there George Duff drew my attention to Lagrange's formula. A reverse application of this formula led eventually to parametric equations for g, via, I suppose, the inverse function of xg. The parametric equations for g are as follows:

$$x = \theta(1-\theta)^3, \tag{10.7}$$

$$xg = \theta(1-2\theta). \tag{10.8}$$

These equations are equivalent to (10.6), as can be verified directly by Lagrange's Theorem. Using them we can substitute for x and g in (10.4) in terms of θ, and then that equation can be solved as an ordinary quadratic for Ψ in terms of y and θ. Thus, using some guesswork, we arrive at a solution of (10.4). The solution can be checked. It is not difficult to show that (10.4) has only one solution for Ψ as a power series in x and y with no negative indices. Hence the guessed-at solution is verified. That is how we solved enumerative equations in the early days of the theory. In [91] I give the general coefficient in Ψ explicitly. The formula is notably more complicated than (10.6).

In [91] I was able to relate the generating functions of the strict and simple triangulations. To do this one uses the fact that a typical strict triangulation can be derived from a uniquely determined simple one by further triangulating some of its faces. Parametric equations were obtained for the generating function of the simple triangulations. An explicit formula for the nth coefficient was obtained, but it was complicated. Later, in [94], I was able to replace it by a simple recursion formula.

In my second "Census" paper [92], I made a similar correlation between the strict triangulations and the general ones. But I went over to the dual form. I found that the number of rooted non-separable planar cubic maps with $2n$ vertices was

$$p_n = \frac{2^n \cdot (3n)!}{(n+1)!(2n+1)!}. \tag{10.9}$$

This is also the number of general rooted triangulations with $2n$ faces.

Let us not forget those map-colouring averages. Having got p_n it seemed reasonable to ask for the number of Tait cycles passing through the root-edge and having s components, summed over all rooted non-separable planar cubic maps of $2n$ vertices. Summing this over s and dividing by p_n we would then get the average number of Tait colourings for such maps of $2n$ vertices. This average would be over rooted maps, and we would of course prefer an average over unrooted ones. But I supposed the two averages would be nearly equal for large n.

In [92] I made my first attack on this problem, confining my attention to the case $s = 1$. I learned that the average number of Hamiltonian circuits for those rooted non-separable cubic maps of $2n$ vertices was

$$h_n = \frac{3 \cdot (2n)! \{(2n+2)!\}^2}{2^{n+2}\{(n+1)!\}^2 (n+2)!(3n)!}. \tag{10.10}$$

An application of Stirling's formula gave the following asymptotic approximation:

$$h_n \sim \frac{8\sqrt{3}}{\sqrt{\pi}} \cdot n^{-\frac{1}{2}} \left(\frac{32}{27}\right)^n. \tag{10.11}$$

I tried to solve the corresponding problem for 3-connected rooted cubic maps, but got no such simple solution. The lesson seemed to be: "Use general triangulations and their duals".

The basic construction was to start with a Hamiltonian circuit, properly rooted, and then to ask in how many ways a 2-connected cubic map could be constructed around it. Presumably we would next need a similar construction for fitting a map to a fixed Tait cycle of s components, a Tait cycle with one edge marked as the root-edge of the map.

Such a Tait cycle would be made up of s disjoint circuits, all of even length. They would partition the sphere into regions I called "bands". Each band would be a connected region bounded either by a single circuit or by two or more disjoint circuits. Constructing the cubic map around the Tait framework would take place in three steps.

(i) Arranging the rooting.
(ii) Partitioning the vertices of each circuit C into two sets U and V, according as to which incident band is to include the third incident edge, the one not in C.
(iii) Joining the chosen vertices across each band B.

Note that in step (ii) the two sets U and V on a circuit C must have even cardinality. For the third edges from U, excluding those with both ends on C, form the support of a coboundary (mod 2) of the graph, and all of them are to receive the same Tait colour.

I concentrated on step (iii). Before an average could be got there would be the awkward task of fitting many bands together. But I would worry about that later.

I called operation (iii) a "slicing" of the band B. It joins the vertices of the appropriate sets U on the bounding circuits of B. It joins them across B so as to divide that region into simply connected polygons. The third "Census" paper sets out to count the possible slicings of B [93].

In this paper the bounding circuits of B are enumerated as $J_1, J_2, \ldots, J_k$. The circuit J_1 is drawn so as to enclose all the other J_i, if there are any. In any case B is represented inside J_1 in a planar drawing, and outside any other J_i. (See Figure 10.6.)

We suppose that on J_i we have $2n_i$ vertices, numbered from 1 to $2n_i$ in cyclic order. This cyclic order should agree with some fixed positive sense of rotation in the plane about discs that are bounded by the circuits J_i and are disjoint from B. With respect to our diagram we can say that the order is clockwise around J_1, and anticlockwise around each other J_i.

In [93] the number of combinatorially distinct slicings of B is written as
$$\gamma(n_1, n_2, \ldots, n_k).$$

The theory of slicings begins with a recursion formula. Consider the edge A drawn across B from the vertex 1 of J_1. If it comes back to J_1 we cut along it and B falls apart into two bands, each representing a simpler slicing problem. If A goes to another circuit J_1 we still cut along it, thus replacing J_1 and J_i by a single new circuit, the external boundary of a new band. We now have to slice the new band. But we consider the problem simplified because the new band has fewer bounding circuits and fewer

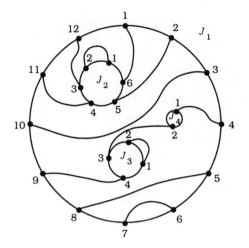

Figure 10.6 A slicing.

effective vertices than the old. In [93] these observations are exploited to obtain the following formula:

$$\gamma(n_1, n_2, \ldots, n_k) = \sum_P \sum_{j=0}^{n_1-1} \gamma(j, P)\gamma(n_1 - j - 1, \overline{P})$$

$$+ \sum_{r=2}^{k} 2n_r \gamma(n_1 + n_r - 1, S - \{J_r\}).$$

(10.12)

Here S is the set of bounding circuits J_i of B with $i > 1$, and P is an arbitrary subset of S. $\bar{P}$ is the complement of P in S. In the symbol $\gamma(j, P)$ we are to replace P by the numbers n_i corresponding to the circuits J_i in P, in proper order. When A returns to J_1 it separates the interior of J_1 into two regions, thereby determining the partition $\{P, \bar{P}\}$ of S.

When $k = 1$ the problem presents no difficulty, being not even new. We adopt the convention that $\gamma(0) = 1$; this convention is to be assumed with formula (10.12). Then $\gamma(n)$ is a Catalan number:

$$\gamma(n) = \frac{(2n)!}{n!(n+1)!}.$$

(10.13)

Using (10.12) it is possible to work from this result to $\gamma(n_1, n_2)$, then to $\gamma(n_1, n_2, n_3)$, and so on. I did this in some simple numerical cases and was then able to guess the following formulae:

$$\gamma(n_1, n_2) = \frac{1}{n_1 + n_2} \cdot \frac{(2n_1)!}{n_1!(n_1-1)!} \cdot \frac{(2n_2)!}{n_2!(n_2-1)!}, \quad (10.14)$$

$$\gamma(n_1, n_2, n_3) = \prod_{i=1}^{3} \frac{(2n_i)!}{n_i!(n_i-1)!}. \quad (10.15)$$

When I had gone along this series a little further I was able to guess a general formula. Writing $2n$ for the total number of vertices involved, that is $2\sum n_i$, I claimed that

$$\gamma(n_1, n_2, \ldots, n_k) = \frac{(n-1)!}{(n-k+2)!} \prod_{i=1}^{k} \frac{(2n_i)!}{n_i!(n_i-1)!}. \quad (10.16)$$

Guessing the formula was one thing, proving it quite another. I expect that my argument in [93] could be much shortened. But such as it is it is the only published proof that I know of.[5]

Once formula (10.16) is accepted we can draw some interesting graph-theoretical consequences from it. We can try contracting the circuits J_1, each with its residual domain outside B, into single vertices. This transforms the sliced band into an Eulerian planar map, a map in which each vertex has even valency. Such a map can have loops but cannot have isthmuses. Accordingly we can deduce from (10.16) the number

$$K(t; q_1, q_2, \ldots, q_s)$$

of rooted Eulerian planar maps with a root-vertex of valency $2t$ and with exactly q_i other vertices of valency $2i$, the maximum valency being $2s$. I did have some difficulty with the transformation, and the formula stated in my commentary in "Selected Papers" is wrong. I now offer the following:

$$K(t; q_1, q_2, \ldots, q_s) = \frac{(n-1)!}{(n-k+2)!} \cdot \frac{(2t)!}{t!(t-1)!} \cdot \prod_{j=1}^{s} \frac{1}{q_j!} \left(\frac{(2j-1)!}{j!(j-1)!} \right)^{q_j}. \quad (10.17)$$

Here n is the number of edges and k is the number of vertices.

Another specialization can be made by putting $n_i = 2$ for each J_i in (10.16), so that all the circuits J_i bound quadrilaterals. We can 3-colour the faces of the resulting bicubic map. The quadrilaterals can be red and the faces inside B alternately green and yellow. It is easy to interpret the red, green and yellow faces as the edges, vertices and faces respectively of

---

[5] It has flaws too. I have spent much time recently convincing myself that they can be corrected.

a planar map. In [94] this construction is used to find the number a_n of rooted planar maps with n edges. The result is

$$a_n = \frac{2 \cdot (2n)! 3^n}{n!(n+2)!}. \tag{10.18}$$

These are general maps, allowing loops and double joins. (But their graphs must be connected, as with all the maps in this book).

In [94] the general rooted planar maps are related to the non-separable ones. The number b_n of rooted non-separable planar maps with n edges is then derived from (10.18), *via* the corresponding generating series. It is found that

$$b_n = \frac{2 \cdot (3n-3)!}{n!(2n-1)!}. \tag{10.19}$$

Going further in this direction I got a formula for the number c_n of rooted 3-connected planar maps. It was not of direct use in calculations, but I was able to deduce from it a reasonably simple recursion formula. I used this to get an asymptotic approximation, and also to calculate c_n up to $n = 25$; c_{25} is 1 932 856 590.

The 3-connected planar maps are of special interest since they correspond to the convex polyhedra. They are also the "c-nets" of the theory of squared rectangles. At the time I was enumerating these maps people in the Netherlands were tabulating them by edge-number in a search for the smallest perfect square [21]. Perhaps it is in this 3-connected case that the enumerative formulae have been most exhaustively checked by numerical computation. Sometimes I would get a letter from Eindhoven urging me to change the formula because it did not agree with the latest count. Then another letter would come saying that the formula had been right after all. Recently the recursion formula for c_n has been simplified by Liu Yanpei [44].

It is possible to enumerate the non-separable planar maps directly, somewhat in the style of the original enumeration of the 3-connected triangulations, and without going back to the theory of slicings. This was done by W.G. Brown in [13]. Later, in [14], Brown and I similarly obtained a formula for the number f_{ij} of rooted non-separable planar maps with $i+1$ vertices and $j+1$ faces, as follows:

$$f_{ij} = \frac{(2i+j-2)!(2j+1-2)!}{i!j!(2i-1)!(2j-1)!}. \tag{10.20}$$

The symmetry between i and j is a consequence of planar duality.

In [95] I tried to relate these maps to the general rooted maps with $i+1$ vetices and $j+1$ faces. I had partial success, getting parametric equations but no simple solution. R.C. Mullin and P.J. Schellenberg, in [47], worked in the other direction, getting results of the same kind about 3-connected maps. For a summary of this enumerative theory of planar maps, up to 1987, see [4].

While I worked on planar enumeration I kept the problem of four-colour averages in mind. I gave up my attempt to find average numbers of Tait cycles with s components by fitting bands together, publishing my meagre results, for $s = 1$ and $s = 2$, in [96], and I went back to an older idea. I now knew the number of general rooted triangulations with $2n$ faces. Why not also find the number of λ-coloured ones, for an arbitrary positive integer λ? Letting T denote a general rooted triangulation, and $K_\lambda(T)$ any λ-coloured one derived from it, one would try to determine the generating function

$$f(z, \lambda) = \sum_T z^{t(T)} \sum_{K_\lambda(T)} 1, \qquad (10.21)$$

that is,

$$f(z, \lambda) = \sum_T z^{t(T)} P(T, \lambda). \qquad (10.22)$$

In this investigation I wrote $t(M)$ for the number of non-root triangular faces in any rooted triangulation or near-triangulation M. The function $P(T, \lambda)$ is the chromatic polynomial of T, or rather of its graph.

It seemed just possible that one might get a functional equation for f by exploiting the recursion formulae for chromatic polynomials.

As in the original enumerative problem I found it necessary to go over to general rooted near-triangulations M and to take account of the root-face valency $m(M)$. The map M was to be non-separable. Its outer face was to be the root-face, and its inner faces were all to be triangles. With one exceptional case the outer face was to be bounded by a polygon that included the root-edge and the root-vertex. The exception was the link-map, having one edge, two vertices and one face. There was essentially only one way of rooting the link-map by the two-arrow method. Its conventional inclusion among the near-triangulations was found to simplify the algebra.

After further study I concluded sadly that I would have to take account of the root-vertex valency $n(M)$ as well. I had to deal with the power series

$$g = g(x, y, z, \lambda) = \sum_M x^{m(M)} y^{n(M)} z^{t(M)} P(M, \lambda). \qquad (10.23)$$

The link-map satisfies $m(M) = 2$, $n(M) = 1$ and $t(M) = 0$. Its chromatic polynomial is $\lambda(\lambda - 1)$. So it makes a contribution

$$x^2 y \lambda(\lambda - 1)$$

to g. As my investigation proceeded I decided that I needed two specializations of g. One is

$$q = q(x, z, \lambda) \tag{10.24}$$

which takes no account of $n(M)$. It is defined as the result of putting $y = 1$ in (10.23). The other,

$$\ell = \ell(y, z, \lambda), \tag{10.25}$$

is the coefficient of x^2 in g. But the function I had set out to find was not one of these; it was $f(z, \lambda)$. That function, I discovered, was very simply related to

$$h(z, \lambda) = \ell(1, z, \lambda). \tag{10.26}$$

So $h(z, \lambda)$ could be defined as above as the result of putting $y = 1$ in ℓ. Equivalently it was the coefficient of x^2 in q.

An equation relating g, q and ℓ is derived in [97]. It is

$$xg = x^2 y \lambda (\lambda - 1) + \lambda^{-1} yzgq + yz(g - x^2 \ell) - x^2 y^2 z \frac{g - q}{y - 1}. \tag{10.27}$$

We can generalize to a non-integral λ, real or complex. The chromatic polynomial is treated in this theory simply as a polynomial in a complex variable λ, with integral coefficients, that satisfies certain recursion formulae. There is nothing in the proof of (10.27) that requires λ to be an integer. But for general λ we can hardly talk of the number of λ-coloured maps. We speak of "chromatic sums" instead.

The subtitle of [97], "The cases $\lambda = 1$ and $\lambda = 2$", may seem a little odd. No near-triangulations can be coloured in 1 colour, and only the link-map can be coloured in 2. Hence $g = 0$ in the case $\lambda = 1$, and $g = 2x^2 y$ in the case $\lambda = 2$. Surely these results are too trivial to justify a paper. Yet when one considers some of the papers that have been submitted and even published ... But I digress. I thought to make this problem non-trivial by asking not for g, q, ℓ and h but for their derivatives g', q', ℓ' and h' with respect to λ at the points $\lambda = 1$ and $\lambda = 2$. Explicit general formulae for the coefficients in these power series are derived in [97]. The method used was the good old one of tabulating numerical results, guessing general formulae and then verifying the guesses by substitution. Unfortunately this method did not work when I substituted $\lambda = 4$ in (10.27) and then tried to solve for the coefficients in $h(z, 4)$.

Besides 1 and 2 there were two other special values of λ for which equations simpler than (10.27) could be found by graph-theoretical arguments. They were $1 + \tau$ and 3, where τ is the golden number $(1 + \sqrt{5})/2$. They are discussed in [98] and [99] respectively. The new equations are between the functions ℓ and h, and they do not involve the variable x. The one in [98] is

$$\tau^3(1 + \tau y)(\ell - \tau^3 y) = y^2 z^2 \ell^2 + \tau^2 y^3 z^2 \frac{\ell - h}{y - 1}, \tag{10.28}$$

and the one for $\lambda = 3$ is

$$(1 - y^2)yz^2\ell^2 - 6(1 - y^2 + z^2y^2)\ell + 36y(1 - y^2) + 6y^3z^2h = 0. \quad (10.29)$$

These equations are not unlike equation (10.4). There is again a quadratic in a function ℓ of two variables, with a function h of one of those variables appearing in the coefficients. At this stage in my studies, in the early seventies, I knew how to solve such equations, and explicit formulae for the coefficients in h are given in [98] and [99]. I suggest a problem for your consideration. Presumably (10.28) and (10.29) can be derived as consequences of the general equation (10.27). But how is this to be done?

My next paper in the "Chromatic sums" series, [100], deals with another special value of λ, a conventional one. It explains that the enumerative theory of near-triangulations can be regarded as the limiting form, as $\lambda \to \infty$, of the theory of their chromatic sums, and discusses this relationship between the two theories.

My fifth paper on chromatic sums, [101], exhibits (10.28) and (10.29) as two members of an infinite series of equations for ℓ for special values of λ. These special values are the "chromatic eigenvalues" scheduled for the next chapter. But [101] has only equations, no solutions. Most of them involve several unknown functions of z, and not h alone. But they can all be shown to have unique solutions.

Not until the early eighties did I make any further real progress. I then discovered, in [102] and [103], how to eliminate ℓ, as well as g and q. The result was a differential equation for $h(z, \lambda)$, valid for all values of λ except 4. The equation, published in [103], runs as follows:

$$(1 - 2\gamma'')(-\lambda\nu^2z^2 + 12z^4 + 20\gamma) + 6z^2\gamma'\gamma'' = 0. \quad (10.30)$$

Here

$$\nu = (4 - \lambda)^{-1}$$

and

$$\gamma = \lambda^{-1}\nu z^4 h + \nu^2 z^2 + 3z^4/2.$$

The prime denotes differentiation with respect to z^2. Note that h is a function of z^2 rather than z. It might be well to change to $t = z^2$ as a new variable.

Even in the case $\lambda = 4$ it is possible, with some care, to derive a differential equation from (10.30). In [103] a function $H = z^4h = t^2h$ is introduced, (10.30) is expressed in terms of H, its t-derivatives, and $4 - \lambda$. Then the limit is taken as $\lambda \to 4$. The resulting equation, in terms of t, is

$$H''(2t + 5H - 3tH') = 48t. \quad (10.31)$$

The coefficient of successive powers of t in H can be calculated from this equation. I am tempted to say that the problem of finding the number of 4-coloured rooted triangulations with $2n$ faces is now solved; it is the appropriate coefficient in the power series H defined as that solution of (10.31) in which the coefficients of t^0 and t^1 are zero. But we still need an asymptotic approximation. For that we can perhaps now look to the theory of differential equations.

I said near the beginning of this chapter that information about averages might be easier to obtain than a proof of the Four Colour Theorem. Yet now we have the Haken–Appel proof, and we still lack an explicit formula or even an asymptotic approximation for our four-colour average.

11

THE CHROMATIC EIGENVALUES

In 1946 an important paper appeared in the *Transactions of the American Mathematical Society*. It was "Chromatic polynomials" by G.D. Birkhoff and D.C. Lewis, referenced here as [8]. It has served as a source and a textbook for students of chromatic polynomials ever since.

Much of the paper is concerned with relating the chromatic polynomial of a planar triangulation T to the chromatic polynomials of smaller triangulations. Actually the paper deals with cubic maps, but in references to it I will put its arguments into their dual forms.

Birkhoff and Lewis consider a circuit C in a triangulation T. It separates that triangulation into two triangulated discs D_1 and D_2. Let λ be a positive integer. The authors observe that each λ-colouring of T can be constructed by combining a λ-colouring of D_1 with one of D_2. Any λ-colourings of D_1 and D_2 that agree on the circuit C can be so combined into a λ-colouring of T.

The authors then introduce the notion of a "constrained chromial". Consider a triangulated disc D bounded by a circuit C. Let

$$\Pi = \{S_1, S_2, \ldots, S_k\}$$

be a partition of the vertex-set $V(C)$ of C into disjoint non-null parts S_i. We are asked to consider the number $Q(D, \Pi, \lambda)$ of colourings of D in λ colours, conforming to the following constraints in C:

I: Two vertices in the same set S_i are to have the same colour;
II: Two vertices from different sets S_i are to have different colours.

It is found that $Q(D, \Pi, \lambda)$ can be written as a polynomial in λ with integral coefficients. It is indeed the chromatic polynomial of a graph, the graph obtained from that of D by identifying all the vertices within each set S_i so as to form a single new vertex V_i, and then joining each two distinct new vertices V_i and V_j by a new edge. It is called a constrained chromial of D, constrained by the partition Π.

Going back to T and its two hemispheres D_1 and D_2 defined by C we can now write

$$P(T, \lambda) = \sum_{\Pi} \frac{Q(D_1, \Pi, \lambda) Q(D_2, \Pi, \lambda)}{\lambda(\lambda-1)\ldots(\lambda-k+1)}, \qquad (11.1)$$

where Π runs through the set of all partitions of $V(C)$. Actually we can restrict the partitions Π to those in which no two consecutive vertices of C belong to the same part S_i. For any two consecutive vertices of C are joined by an edge of C, and must therefore receive different colours in every λ-colouring of T, of D_1 and of D_2. In the rest of this chapter I will assume the partitions of a circuit C to be so restricted.

Having found equation (11.1) Birkhoff and Lewis could rejoice in being able to express the chromial of T in terms of constrained chromials of the two discs defined by C. In non-trivial cases C could be chosen so that D_1 and D_2 had smaller graphs than T.

They next asked: "How shall we express a constrained chromial $Q(D, \Pi, \lambda)$ of a disc D in terms of the chromials of triangulations T_D associated with D?" If each constrained chromial in equation (11.1) were so expressed there would result an expression for $P(T, \lambda)$ in terms of the chromials of smaller triangulations. Surely such an expression could be used to prove inductively theorems about chromatic polynomials? So Birkhoff and Lewis proposed to construct a number of triangulations from their triangulated disc D by making joins and identifications across the outer face. The chromials of these triangulations they described as "free chromials" of D. They now asked: "How shall we express the constrained chromials of D in terms of its free chromials?"

I think I can best explain their procedure by giving an example, an example that presents the simplest non-trivial case. We take D to be bounded by a quadrilateral C whose vertices are numbered 1, 2, 3 and 4 in cyclic order, as in Figure 11.1.

The relevant partitions of $V(C)$ can be written as

$$(1, 2, 3, 4), \ (13, 2, 4), \ (1, 24, 3), \ldots, (13, 24).$$

In each case the parts are separated by commas. Thus the second partition has three parts. One consists of the vertices 1 and 3, and each of the others has one vertex only. In the last partition there are two parts only, one with vertices 1 and 3 and the other with vertices 2 and 4. To these four partitions of $V(C)$ there correspond four constrained chromials of D, which I write respectively as

$$\{1, 2, 3, 4\}, \ \{13, 2, 4\}, \ \{1, 24, 3\}, \ \{13, 24\}.$$

We can turn D into a triangulation by joining vertices 1 and 3 across the outer face. Call this triangulation Z_1. Then $P(Z_1, \lambda)$ is one of the free chromials of D. Clearly it is the sum of those constrained chromials of D that have vertices 1 and 3 in different parts. Thus

$$P(Z_1, \lambda) = \{1, 2, 3, 4\} + \{1, 24, 3\}. \tag{11.2}$$

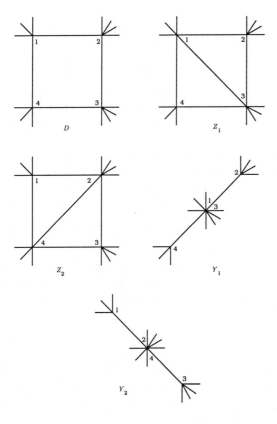

Figure 11.1 Maps defining free chromials.

Similarly we can form from D a triangulation Z_2 by joining vertices 2 and 4. We then have

$$P(Z_2, \lambda) = \{1,2,3,4\} + \{13,2,4\}. \tag{11.3}$$

Another possibility is to close up the outer face, identifying vertex 1 with vertex 3, edge 12 with edge 32, and edge 14 with edge 34. Call the resulting map Y_1. It is a true triangulation unless it contains a loop, originally an edge joining vertices 1 and 3 in D. If it does have a loop its chromatic polynomial must be zero, and the loop will therefore cause us no trouble. In any case $P(Y_1, \lambda)$ is ranked with the free chromials of D. It is the sum of those constrained chromials of D for which vertices 1 and 3 belong to the same part. Thus

$$P(Y_1, \lambda) = \{13,2,4\} + \{13,24\}. \tag{11.4}$$

A similar identification, putting vertices 2 and 4 together, gives us a map Y_2. The corresponding free chromial $P(Y_2, \lambda)$ satisfies

$$P(Y_2, \lambda) = \{1, 24, 3\} + \{13, 24\}. \tag{11.5}$$

Equations (11.2) to (11.5) express four free chromials in terms of the four constrained ones. This of course is the wrong way round; Birkhoff and Lewis wanted the constrained chromials in terms of the free. We do have four equations for the four "unknown" constrained chromials, and we might hope to solve them. However an attempt to do this will fail because the four expressions on the right are not linearly independent. If we add equations (11.2) and (11.4) and then subtract (11.3) and (11.5) the constrained chromials all cancel out and we are left with the identity

$$P(Z_1, \lambda) - P(Z_2, \lambda) + P(Y_1, \lambda) - P(Y_2, \lambda) = 0. \tag{11.6}$$

This is a well-known recursion formula.

At this stage Birkhoff and Lewis present us with a new equation of mysterious provenance, one that involves the constrained chromials only. They give two proofs of it. It runs as follows:

$$\{1, 2, 3, 4\} + (\lambda - 1)(\lambda - 2)\{13, 24\} \tag{11.7}$$
$$= (\lambda - 3)\{13, 2, 4\} + (\lambda - 3)\{1, 24, 3\}.$$

Birkhoff and Lewis now have their four independent linear equations and can solve for the constrained chromials. They find for example that

$$(\lambda^2 - 3\lambda + 1)\{1, 2, 3, 4\} = \lambda(\lambda - 3)P(Z_1, \lambda) \tag{11.8}$$
$$+ (\lambda - 3)P(Y_1, \lambda) - (\lambda - 1)(\lambda - 3)P(Y_2, \lambda).$$

In this equation it has been arranged that the coefficients on the right are polynomials in λ. Evidently it has been necessary to rationalize by multiplying throughout by $\lambda^2 - 3\lambda + 1$, a polynomial deriving from the determinant of the system of equations.

Birkhoff and Lewis, at this stage, were well placed for dealing with triangulations having a separating quadrilateral. They went on to treat the case of a 5-circuit C in the same way. The story was much the same, but with more equations and more chromials. Again they initially had not enough independent equations. Again they remedied the deficiency by pulling a new equation out of a hat, a linear equation between constrained chromials only. This time the new equation was

$$\{1, 2, 3, 4, 5\} + (\lambda - 3)(\lambda - 4)\{3, 14, 25\} \tag{11.9}$$
$$= (\lambda - 4)\{1, 25, 3, 4\} + (\lambda - 4)\{14, 2, 3, 4\},$$

in an obvious extension of the preceding notation. To this we should add the companion equations derived by cyclic permutations of the sequence

$(1, 2, 3, 4, 5)$. Again the authors gave a special proof for the new equation. Again they deduced equations giving the constrained chromials in terms of free ones, and again the multiplier $\lambda^2 - 3\lambda + 1$ appeared on the left. But by now they had decided that their equations looked simpler in terms of the variable $u = \lambda - 3$, so they wrote the multiplier as $u^2 + 3u + 1$.

In the paper there is some discussion of 6-circuits and longer circuits. The hope is expressed that a complete theory for separating circuits of any length may make possible a solution of the Four Colour Problem. But for a complete theory of the 6-circuit we have to turn to [33], a later paper of D.C. Lewis and Dick Wick Hall. The procedure is essentially the same, but the multiplier on the left of the final equations is more complicated. It is $6f_1f_2$, where

$$f_1 = u^2 + 3u + 1, \quad f_2 = u^3 + 4u^2 + 3u - 1. \tag{11.10}$$

My own interest in chromatic polynomials dates from the late forties. I had a vague idea that the zeros of chromatic polynomials ought to be studied and calculated for some of the planar triangulations. I believed that whenever possible the variable λ of chromatic polynomials should be allowed to take any real or complex value. It should be noted that this can be done in the equations of Birkhoff and Lewis. They are equations between polynomials in λ that are valid for infinitely many values of λ, namely the positive integers. Hence they must be polynomial identities. So I was delighted when in 1965 a new paper appeared, by D.W. Hall and co-workers, giving all the zeros of the chromatic polynomial for one particular triangulation. The chromial is described in the paper as that of the truncated icosahedron, a cubic map. But in the dual terminology of this lecture it is the chromial of the map derived from the regular dodecahedron by erecting a five-sided pyramid on each of the 12 faces. I was told that the necessary calculations by Hall and his students had extended over 15 years. They worked recursively through the chromials of some 900 intermediate maps. At the end, it would seem, the Computer Age dawned in their department and they were able to publish a list of all the 32 zeros [34].

There were the trivial zeros 0, 1, 2 and 3, and there were four real non-integral ones. All the others were complex.

I told my colleague Gerald Berman of my hopes for more calculations of zeros. He said he had just written a new programme for calculating zeros of polynomials and he wanted to try it out. And Ruth Bari contributed a copy of her thesis, in which there is a list of the chromials of 100 maps, of the kind considered in Four Colour Theory. Berman found the zeros of all these. Later D.W. Hall showed us his list of 900 chromials and Berman found the zeros of them too. All these were, after dualizing, chromials of 3-connected planar triangulations.

We noticed one regularity. All, or nearly all, of these maps had a zero nearly equal to 1.618. It soon appeared that the more complicated maps had zeros agreeing to 5, 6 or 7 places with

$$1 + \tau = \tau^2,$$

where τ is the golden number $(1 + \sqrt{5})/2$.

This observation was recorded in a paper, [5], published in 1969. Before publication I wrote to Dick Wick Hall asking if he had noticed that one of his real zeros, for the truncated icosahedron, was to 8 places of decimals $1 + \tau$. He replied "No". But he pointed out that $1 + \tau$ was a zero of the Birkhoff–Lewis multiplier $\lambda^2 - 3\lambda + 1$, the one that appears in our equation (11.8) and that is denoted by f_1 in the paper of Lewis and Hall. Moreover a real zero of his multiplier f_2, a cubic, agreed to 5 places of decimals with another of the real zeros of his truncated icosahedron.

Not long after this I was able to prove some theorems about the values of chromatic polynomials at $\lambda = \tau + 1$ for planar triangulations. One of them states that for any non-separable planar triangulation T with k vertices,

$$|P(T, \tau + 1)| \leq \tau^{5-k}. \tag{11.11}$$

This was published in 1970, in [104]. It seemed to explain the tendency of these chromials to have zeros near $\lambda = 1 + \tau$. It did not prove that such a zero must exist in every case. This was fortunate, for we now know of triangulations without such zeros. There was an interesting auxiliary theorem saying that $1 + \tau$ itself could never be a zero.

I found it amusing to search among Hall's 900 for different triangulations with the same chromial, or for pairs of triangulations whose chromial-difference, though non-zero, was of unusually low degree. Attempts to factorize such differences uncovered a regularity. Some of them divided by $\lambda - \tau^2$ and the conjugate factor $\lambda - \tau^{*2}$. And when this happened they divided also by $\lambda - \tau\sqrt{5}$ and $\lambda - \tau^*\sqrt{5}$. Note that $\tau\sqrt{5} = \tau^2 + 1 = \tau + 2$. After much fruitless work a possible explanation of this effect occurred to me. It might be that, for any triangulation T, the number $P(T, \tau\sqrt{5})$ could be expressed as a function, independent of T, of $P(T, \tau^2)$. A few points plotted on squared paper indicated that this would be the square function, multiplied by some constant. In the end I decided that

$$P(T, \tau\sqrt{5}) = \sqrt{5} \cdot \tau^{3(k-3)} \cdot P^2(T, \tau^2). \tag{11.12}$$

This result was published in [105]. The proof given there is marred by a fallacious extension beyond the realm of triangulations. But for triangulations it is valid.

Formula (11.12) is called the Golden Identity. Sometimes people find it difficult to believe at first. Dick Wick Hall told me he had not done so until he had selected a moderately complicated triangulation and computed both sides of the equation to 14 decimal places.

The Golden Identity has one interesting consequence. It is known, as I remarked above, that $P(T, \tau^2)$ is never zero. Hence $P(T, \tau\sqrt{5})$ is always positive. So $P(T, \lambda)$ is positive at $\lambda = 3.618$ approximately. According to Haken and Appel it is positive also at $\lambda = 4$. But among Hall's 900 there are examples with zeros between these two values of λ. It is proved in [8] that $P(T, \lambda)$ is always positive for $\lambda \geq 5$. There is no known case of a zero between 4 and 5. Nor have we any proof that such a zero is impossible.

As for the zero of f_2 we did find some clustering of zeros of chromials around it. But the effect was much less marked than for the "golden root" $\tau^2 = \tau + 1$. However at Waterloo it became known as the "silver root".

At about this time Sami Beraha was urging us to look very carefully at all values of λ of the form

$$B_n = 2 + 2\cos(2\pi/n), \qquad (11.13)$$

with n a positive integer. Evidence was accumulating that they had some special significance in the theory of chromials. Consider the initial values of n. We have $B_1 = 4$, and no value of λ has received more attention from mathematicians. $B_2 = 0, B_3 = 1$ and $B_4 = 2$, and these values of λ are zeros of the chromials of all planar triangulations. B_5 is the golden root, B_6 is the integer 3, and B_7 is the silver root. We cannot deny the significance of the number 3 in this context; it is a zero of the chromial of every triangulation that is not Eulerian.

B_8 is $2+\sqrt{2}$. It does agree to 2 places with a zero of the truncated icosahedron. But there is more than that. D.W. Hall has carried the Birkhoff–Lewis theory, in a simplified form, up to the 7-circuit [32]. He deals not with single chromials but with sums of symmetrically equivalent ones. Again he gets a polynomial multiplier on the left, and this multiplier divides by $\lambda - B_8$.

All I claim here for B_9 is that it agrees to one place of decimals with the last real root of the truncated icosahedron. But B_{10} is $\tau\sqrt{5} = \tau + 2$, and it is glorified by the Golden Identity.

It is natural to conjecture that the Birkhoff–Lewis theory of the m-circuit always gives a multiplier on the left that divides by $\lambda - B_{m+1}$. Equivalently we can conjecture that the matrix of the final system of linear equations has a zero determinant when λ is B_{m+1}. One longs for a simplification of the Birkhoff–Lewis theory that would make this clear.

Having come back to the free–constrained equations let me comment again on the constrained–constrained ones, those that Birkhoff and Lewis pulled out of their hat. They do give a partial theory of these equations in their paper, but I confess that I was never able to read right through it and understand it clearly.

When Frank Bernhart was at Waterloo he and I studied those equations. When discussing partitions $\Pi = \{S_1, S_2, \ldots, S_k\}$ of $V(C)$ we made a distinction between "planar" and "non-planar" ones. In a non-planar partition there are two distinct parts S_i and S_j such that two vertices of S_i separate two vertices of S_j in the circuit C. In a planar partition no two parts S_i and S_j are so related. We decided that the basic hat-equations expressed the constrained chromials for the non-planar partitions in terms of those for the planar ones. We therefore called them the "flattening equations". Bernhart developed a very pleasing theory of them. It has sometimes been the subject of lectures but it has not yet been published.

Let me next mention some more theorems about the golden root. There is one that I call the Vertex-elimination Theorem. It relates the chromial of a triangulation T to that of the map T_ν obtained from T by deleting a vertex ν and its incident edges. It asserts that

$$P(T, \tau^2) = (-1)^m \cdot \tau^{1-m} \cdot P(T_\nu, \tau^2), \qquad (11.14)$$

where m is the valency of ν. But the theorem can be made more general. It is not necessary for the map T to be a triangulation, only that all the faces incident with ν shall be triangles. For example we can apply the theorem to the deletion of the twelve 5-valent vertices of the dualized truncated icosahedron, one by one. We then relate the face-chromial of the truncated icosahedron to the vertex-chromial of the regular dodecahedron (at $\lambda = \tau^2$).

When I heard from Norman Biggs that a student, David Sands, had calculated the vertex-chromial of the dodecahedron I advised that Sands should use the Vertex-elimination Theorem to calculate the face-chromial of the truncated icosahedron at $\lambda = \tau^2$. I thought that agreement with the published chromial would be convincing confirmation of the accuracy of the latter. Dick Wick Hall had told me that he sometimes had qualms about the published result. In 15 years of calculation there must be some possibility of making an error. It had been a great comfort to him to know that his chromial had real zeros making sense in terms of the Beraha numbers.

Sands took the advice and reported eventual agreement. He did not get initial agreement, but he located the error in his own calculations and not in Hall's. Another powerful check would be to verify that the published chromial satisfies the Golden Identity. I do not know if this has been done.

Another of the "golden" theorems can be expressed in terms of the Y's and Z's of equation (11.6). It is an equation much like that one in form but with different coefficients. It can be written as follows:

$$P(Z_1, \tau^2) + P(Z_2, \tau^2) = \tau^{-3}\{P(Y_1, \tau^2) + P(Y_2, \tau^2)\}. \qquad (11.15)$$

This equation can be taken as the basic one in the theory of golden chromials. We can then explain the special character of that theory by

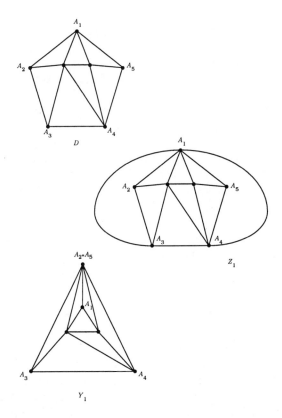

Figure 11.2 Seeking a special identity with $\lambda = 3$.

saying that at $\lambda = \tau^2$ the chromials of triangulations satisfy two linear recursions instead of one.

When we are talking of the golden root $\tau^2 = \tau + 1$ we should not overlook its algebraic conjugate $r^* + 1 = (3 - \sqrt{5})/2$. That is essential at one crucial point in the golden theory. It is easy to show that the chromial of a connected graph G, without loops, has no zero between 0 and 1. Hence $P(T, \tau^* + 1)$ is non-zero for any triangulation T. It follows by algebraic arguments that $P(T, \tau+1)$ must be non-zero too, as I have already claimed. Algebraic identities like the Golden Identity are true for $\tau + 1$ if they are true for $r^* + 1$. But with formulae like (11.11), involving an inequality, we must be very careful in any transformation to a conjugate root.

Similarly any other Beraha number can be expected to share some of its chromatic properties with its algebraic conjugates.

If we take the equations of Lewis and Hall in [33], giving constrained chromials in terms of free, and substitute the silver root $\lambda = B_7$, the left sides vanish. Each equation reduces to an identity, and each to the same identity. It relates the chromials at B_7 of several triangulations simply derived from the triangulated disc D. It is a special silver identity, valid only at B_7 and its algebraic conjugates. It should permit the construction of a silver theory just as the analogous equation (11.15) provides the basis for a golden one. But as yet no one has seen how to exploit it; it involves too many different triangulations.

One supposes that there is one such special identity for each Beraha number B_n, associated with a circuit of $n-1$ edges. I was myself able to find such an identity for B_6, which is 3, and the 5-circuit. Consider a disc D bounded by a pentagon, as sketched in Figure 11.2. The vertices of the pentagon are denoted by A_1, A_2, A_3, A_4 and A_5, in cyclic order, and the suffixes are regarded as residues mod 5.

For each suffix i let a triangulation Z_i be formed by joining A_i to A_{i+2} and A_{i+3} across the outer face of D. Let also Y_i be formed by identifying vertices A_{i-1} and A_{i+1}, with a corresponding identification of the segments joining them to vertex A_i. I found that

$$P(Y_i, 3) = P(Z_{i+2}, 3) + P(Z_{i+3}, 3) \tag{11.16}$$

for each suffix i (see [99]).

There is a minor mystery here. Why does not $(\lambda - 3)$ occur as a multiplier in the Birkhoff–Lewis equations for the pentagon? I have stated in lectures that B_6 bears the same relation to the pentagon as does B_5 to the quadrilateral or B_7 to the hexagon. Sometimes a disciple of Birkhoff and Lewis has rebuked me for this, saying that it is not so written in the Great Paper. However, he would point out, there are equations in that paper from which my (11.16) can be got by substitution.

In the previous chapter I told you something about the theory of chromatic sums. I told you how, for a general λ I got my equation (10.27) between the generating functions g, q and ℓ, all of them sums over 2-connected rooted near-triangulations. I stated that for $\lambda = \tau^2$ and $\lambda = 3$ there were the simpler equations (10.26) and (10.27) involving only ℓ and its specialization h. The basic reason for these simplifications is that for these values of λ there are extra identities to exploit, (11.15) in the former case and (11.16) in the latter. (See [98] and [99].) Notice how the theory of chromatic sums in the seventies had its successes, apart from the conventional value ∞, only at the simpler Beraha numbers, $B_3 = 1, B_4 = 2, B_5 = \tau^2$ and $B_6 = 3$. The case $\lambda = 3$ is in a way trivial since it reduces to a straightforward enumeration problem, the enumeration of rooted non-separable Eulerian triangulations.

Successes at B_5 and B_6 inspired me to make a great effort to eliminate $g(x, y, z, \lambda)$ from the general equation. An account of the results is given in [101]. It appeared that the elimination could be carried out at some special values of λ, these being the Beraha numbers, the "chromatic eigenvalues" of the title. The general equation for B_n is written in [101] in a somewhat lengthy and repellent form. But the essentials can be stated as follows.

For B_7 and B_8 a cubic equation for $\ell(y, z, \lambda)$ can be given, involving two unknown functions $h_1(z, \lambda)$ and $h_2(z, \lambda)$ of z alone. For B_9 and B_{10} a quartic equation for ℓ can be given, involving three unknown functions of z. And so on. There is in [101] an argument to show that the equation, for each n, contains enough information to fix the coefficients in ℓ and the associated unknown functions of z. But no solution, for any n, is given. I could claim to have demonstrated that all the Beraha numbers are significant in the theory of chromatic polynomials of triangulations.

The theory is, I hope, somewhat clarified in [106], and parametric solutions for the unknown functions of z are given in [102]. Usually there are inconveniently many parameters. However a formal parametric solution can be given for each Beraha number. Beraha pointed out that the limit of B_n, as $n \to \infty$, is 4. He suggested that for this reason alone 4 must be a number of exceptional significance in the theory of planar chromials.

In [103] the parametric equations are used in a derivation of a differential equation for the chromatic sum $h(z, \lambda)$. They prove it only at the Beraha numbers. But an argument is then given that extends the validity of the differential equation to all real or complex values of λ. The differential equation for a general λ appears in the preceding chapter as (10.30), and a derived equation for the case $\lambda = 4$ is stated in (10.31).

Perhaps there is a shorter path to the differential equation that avoids the theory of Beraha numbers. But these numbers are of basic importance in the theory of chromatic sums, and their role in the general theory of chromials ought to be clarified.

12

IN CONCLUSION

I write this commentary on the preceding chapters some years after the lectures were delivered. For some chapters, especially the first, there are things I would like to add, since the amount of information that can be imparted in a single lecture is limited.

Take the first chapter. I need not add to its account of the Four encountering the problem, of our constructing a small catalogue of perfect rectangles, of our noticing the relevance of Kirchhoff's Laws, and of our exploitation of an electrical textbook. Surely an exemplary application of the Scientific Method.

Chapter 1 goes on to the theory of rotors and I take this opportunity to tell again how we arrived at that theory. It happened when our catalogue of perfect rectangles had covered the 9th and 10th orders, had included all or most of the 11th and 12th, and was encroaching upon the 13th.

Among our specimens of the 13th order was a rectangle whose p-net was the cube-graph with one face-diagonal. It attracted our special attention because its reduction was 6, which struck us as unprecedentedly large. Its reduced sides were 75 and 112, and the sides of its constituent squares ranged from 3 to 42.

Brooks was impressed. On vacation he drew the rectangle on cardboard and cut it into its constituent squares. He now had a puzzle; he could challenge someone to fit those squares together to make a rectangle.

On his return to Cambridge he told us how he had challenged his mother, how he had watched with amusement as she started with a sequence of wrong moves, and of his astonishment when she completed a rectangle that was different from the one he had cut up. That was the key discovery of the whole research. The electrical networks (or "p-nets") are shown in Figure 12.1, our own on the left and Mrs Brooks' on the right.

We now had two perfect rectangles of the same shape. They had the same reduced elements (though not the same reduction). We would have preferred them with all different ones, making possible a perfect square. But you can't have everything at once. After a study of the graphs of Figure 12.1 we were able to explain the phenomenon in electrical terms. The graph on the left falls into three parts meeting each other only in the A's. Two of them have a symmetry under which the three A's are mutually equivalent. We inferred that the vertices P and P' were at the

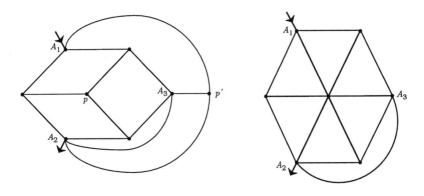

Figure 12.1 Mrs Broook's discovery.

same potential, the average of the potentials of the A's. Hence P and P' could be put together without causing a spark—I mean, without altering the currents in the wires. That identification of P and P' produces the p-net on the right of the figure.

I resolved to construct some more examples of this "isomerism" (a chemical term). I would need some simple graphs of threefold symmetry to replace one or both of the two symmetrical parts of Figure 12.1. It was natural to choose a simple rotor. The one I chose was the one you can obtain from that of Figure 1.6 by shrinking the central triangle to a single vertex. Substituting it for the larger symmetrical part of Figure 12.1 I calculated another pair of isomers. It then seemed to me that the job would not be properly done until I had put in the rotor the other way and calculated the corresponding isomers. I was startled to see that all four rectangles had the same shape and size, and delighted to notice that rectangles from different isomeric pairs had only a few elements in common.

Thus was discovered the rotor–stator effect, though its proof was long delayed. It was not deducible from the isomeric effect and yet a student of isomers could hardly avoid discovering it.

Chapter 1 tells how rotor–stator theory led to the construction of a perfect square. It does not tell how the first attempts failed, producing pairs of perfect rectangles of the same shape but with too many elements in common. I confess I was disheartened. But the other members of the Four thought we had been using rotors that were too simple; the secret was to use more complicated ones. Try something really complicated! So it came to pass that Smith and Stone sat down together and calculated a perfect square of the 69th order. Taking it in triumph to Brooks' room they entered and said: "Look at this". And Brooks, having looked, brought out a paper of his own and said: "Look at this". Another perfect square of the 69th order! Priority is uncertain.

You can read in [11] of how Stone later used the rotor–stator method to exhibit an infinite family of perfect squares. He found that this family contained an infinite sequence of "totally different" perfect squares. "Totally different" meant that no two members of the sequence, brought to the same size, had a pair of congruent constituent squares. He deduced that any rectangle with commensurable sides could be squared perfectly. You cut it up into equal squares in the obvious way and then fill up all but one of them with members of the totally different sequence.

This little story might be held to show the importance of luck in scientific research. How fortunate that Brooks chose the only rectangle in our catalogue that happened to have an isomer! But one can point out that Mrs Brooks' isomeric pair would have been discovered anyway, in the process of completing the 13th order. Yet perhaps the Four would not have bothered to complete the cataloguing of that order without the stimulus of a serendipitous discovery or two?

Let us also note an example of the common phenomenon of two separated researchers getting similar results independently and at near enough the same time. While we worked in Cambridge there was R.P. Sprague, seeking the perfect square in Berlin. Soon he was to be on the far side of battle-lines. His was the first perfect square to be published [55]. He even got in first with an infinite family of perfect squares and a proof that commensurable-sided rectangles are perfectible [56]. And all this by the empirical method, which starts with a catalogue of rectangles and observes how some can be fitted together to make still more interesting figures.

I cannot cease from commenting on Chapter 1 and reference [11] until I have told you of our excursion into the Theory of Numbers. We wrote C for the complexity (or tree-number) of a p-net. We also wrote $[ab.cd]$ for the full potential drop from c to d when a is the positive pole and b the negative. We knew that $[ab.cd] = [cd.ab]$. By determinant theory we got the intriguing congruence

$$[ab.ab]\,[cd.ef] \equiv [ab.cd]\,[ab.ef] \mod C. \qquad (12.1)$$

(See equations 2.33 and 2.34 of [11].) We note that $[ab.ab]$ is the vertical side of the associated rectangle, as C is the horizontal side.

From equation (12.1), with $ef = cd$, we could infer that if $[ab.ab]$ had highest common factor m with C then $[ab.cd]$ would divide by the "upper square root" of m, that is, the product of the square-free part and the square root of the square part. The rectangle would then have that upper square root in its reduction. We marvelled at what this implied for a perfect square, for which the highest common factor is C.

IN CONCLUSION

It was Stone who proved our main theorem on imperfect rectangles. An "imperfection" is a pair of congruent squares in a squared rectangle, that is, a pair of wires with equal currents in the p-net. An imperfection is "trivial" if the two wires have the same ends, or if they constitute a bond not separating the poles. Otherwise the imperfection is "non-trivial". The theorem asserts that all non-trivially imperfect rectangles are reducible, that is, have reductions exceeding 1. The proof in [11] is short, but it was hard to find. It uses the above congruence.

I liked one partial converse of the theorem, telling me that if C and $[ab.ab]$ were coprime the corresponding rectangle would be perfect. (Trivial imperfections were excluded by our normal choice of networks). In particular if C were prime for some c-net N then every squared rectangle derived from N, by whatever choice of poles, would be perfect.

I sometimes used the congruence of equation (12.1) to extend the catalogue. Suppose for example that I had calculated the full flow in a network N for poles a and b. Suppose I had noticed that C and $[ab.ab]$ were coprime. Then I could find the flow with poles c and d as follows. I would determine any $[cd.ef]$ that I needed, mod C, by solving congruence (12.1). I would normally choose e and f to be adjacent vertices, so that $[cd.ef]$ would be the current in a wire. Not all the currents for poles c and d need be calculated in this way; only enough for the others to be inferred by direct applications of Kirchhoff's equations. I found it easy to convert mod C solutions into true solutions. I recall that in this activity I sometimes turned for help to the theory of quadratic residues, but I do not remember how or why.

Extensive work on squared rectangles has been done at Eindhoven. It began with C.J. Bouwkamp during the Second World War and has been continued by himself, A.J.W. Duijvestijn and others. Using computers they extended the catalogue of squared rectangles so far that perfect squares began to appear in it. From their work we know that there is a simple perfect square of order 21, but no other such square of order less that 22. A "simple" perfect square or rectangle is one that contains no smaller perfect rectangle. The others are "compound". We now know that there is no compound perfect square either of order less than 22.

Another worker is T.H. Willcocks of Bristol, England, who long held the low-order record for perfect squares. His square, a compound one, is of the 24th order. He first published it in a magazine of "fairy chess" [119].

Then there is Jasper D. Skinner II, an enthusiastic dissector of squares into unequal squares, or alternatively into unequal isosceles right-angled triangles. He works in Lincoln, Nebraska, where he has found a multitude of perfect squares in computer searches. Even isomeric perfect squares have been found. See *Squared squares, who's who and what's what* by J.D. Skinner II (1993).

My commentary on Chapter 1 is now complete. I shall have less to say about each of the other chapters, for in them I did not have to leave out so much material that had been of direct concern to me.

I see nothing that need be added to Chapter 2 beyond a recommendation to read the chapter on Kozyrev–Grinberg theory in Ross Honsberger's *Mathematical Gems*, Vol. 1 (Mathematical Association of America, 1973). But perhaps the reader will permit the trivial remark that the Hamiltonicity of 4-connected planar maps makes them 4-colourable, as regards faces.

I find nothing more to add until I come to Chapter 7, "Graphs on spheres". For its purposes a sphere is equivalent to a closed plane, and in practice the chapter reduces to a study of plane graphs. But a theorem that does not distinguish between the infinite face and the others is really a theorem about graphs on spheres.

I would like to tell of one more theorem in connection with Chapter 7, a theorem that is strictly planar. It arose out of some speculations as to what would be the best way to draw a 3-connected graph in the plane. I knew it could be done with straight edges and I thought it could be done with convex finite faces too. I proposed to draw the border of the outer face as that of a convex polygon. I would then try to position the remaining vertices within that polygon so that the sum of the squares of the distances between pairs of adjacent vertices was a minimum. I discovered that this minimum was attained when each inner vertex was at the centroid of its neighbours (including any on the outer pentagon). This condition gave two sets of linear equations, one for the x-coordinates and one for the y's. Each set of equations was like Kirchhoff's for a p-net. The x's for example played the part of electrical potentials, the edges were unit conductances, and the outer vertices were the poles. True, there were now more than two poles, but that was a detail. So the two sets of equations could be solved in manners familiar to me to give what I called a "barycentric representation" of the 3-connected planar graph.

There was a most intriguing theory that eventually assured me that the inner vertices occupied distinct points, that no edges crossed, and that all the internal faces were convex. I do not claim that this barycentric solution is the best way to draw a graph. But at least, when the outer frame is specified, the solution is given uniquely by a set of linear equations; it is not a matter of trial and error. An obvious disadvantage is that usually some faces are found to be very much smaller than others. The theory is set out in my paper "How to draw a graph" (*Proc. London Math. Soc.*, **52** (1963), 743–767). Presumably the theory would work equally well with an arbitrary set of positive conductances, not just unit ones, and a judicious choice of conductances might make the faces less dissimilar in area. But I have made no study of that possible refinement.

I tried to get an analogous representation on the sphere, one that did not distinguish an "outer" face, but no results of significance emerged. I would explain the problem as follows: Let the 3-connected planar graph be made of perfect elastic, the restoring force in any edge being proportional to the edge-length. Let it be stretched over a smooth sphere. Can it be in equilibrium there, or must it fly off?

Coming to Chapter 8, I take the opportunity to mention another paper of mine on matroid theory. It is one of those obtained by taking a known theorem about graphs and generalizing it to matroids. A "wheel" is the graph obtained from a circuit and one extra vertex by joining that extra vertex (the "hub") to every vertex of the circuit (the "rim"). The graph-theorem to be generalized says that if a 3-connected graph, as I define such things, is not a wheel then it can be changed into a smaller 3-connected graph by deleting or contracting a suitable edge. See my paper "A theory of 3-connected graphs" (*Konink. Nederl. Akad. van W., Proc.* **64** (1961), 441–455).

The theorem can be specialized to planar graphs, and that special case has been used by C.J. Bouwkamp and his colleagues in the cataloguing of c-nets, a stage in the cataloguing of squared rectangles. The planar case was discovered and used by T.P. Kirkman in the 1850s. But I did not know that in 1961. See Norman Biggs' paper "T.P. Kirkman. Mathematician" (*Bull. London Math. Soc.*, **13** (1981), 97–120).

The statement of the matroid theorem is similar but there are two exceptional matroids, each related to the graphic wheel. One is the circuit-matroid of a wheel, and I still call it a "wheel". The other I call a "whirl". It is derived from the corresponding wheel as follows. We cease to count the rim as a matroid-circuit. We then recognize as many new circuits as there are spokes to the wheel. Each is the rim of the wheel plus one spoke. See my paper "Connectivity in Matroids" (*Can. J. Math.*, **18** (1966), 1301–1324).

I pass over Chapter 9 on reconstruction with a repetition of the comment that the theory badly needs some new idea.

I should bring Chapter 11 up to date by mentioning that since I wrote it I have done some research on the Birkhoff–Lewis equations. First I had an enjoyable time simplifying and generalizing them. I restricted the constrained chromials to those defined by planar partitions. I restricted free chromials to those obtained by identifications of outer vertices, the sets to be identified being given by the parts of a planar partition. I abandoned the restriction to triangulations. I abandoned the requirement of an outer circuit. Instead I required only that some "outer" vertices were to lie on a geometrical circle while the rest of the graph lay inside it. I even managed to work with a class of edgeless graphs, mere sets of vertices with specified identifications. In the domain of edgeless graphs one could obtain formulae for the Birkhoff–Lewis equations, and these formulae could be generalized to the edged graphs by induction. The induction was by deletion and contraction of edges.

Linear equations for n outer vertices were thus obtained. Theoretically they could be solved by Cramer's Rule for constrained chromials in terms of free. I say "theoretically" because the solution required the inversion of a big matrix, one that I now call the "matrix of chromatic joins". It was desirable to find the determinant of this matrix; among its factors should be "Beraha" polynomials such as had occurred in solutions of older equations for small values of n. I had many discussions with my Waterloo colleague Dan Younger about this matrix. He and some colleagues evaluated the determinant by computer for enough small values of n to be able to conjecture a general formula. And that conjectured formula was a product of powers of Beraha polynomials. After many false starts I found a way to prove it. See "The matrix of chromatic joins" (*J. Combinatorial Theory, Ser.* B, **57** (1993), 269–288).

The matrix has since been inverted by Younger's student R. Dahab, as explained in his Ph D thesis.

BIBLIOGRAPHY

1. L. Auslander and H.M. Trent (1959). Incidence matrices and linear graphs. *J. Math. Mech.*, **8**, 827–835.
2. W.W. Rouse Ball and H.S.M. Coxeter (1987). *Mathematical recreations and essays*, 13th edition. New York.
3. K.A. Berman (1980). A proof of Tutte's trinity theorem and a new determinant formula. *SIAM J. Alg. Disc. Math.*, **1**, 64–69.
4. E.A. Bender (1987). The number of three-dimensional convex polyhedra. *Amer. Math. Monthly*, **94**, 7–21.
5. G. Berman and W.T. Tutte (1969). The golden root of a chromatic polynomial. *J. Comb. Theory*, **6**, 301–302.
6. N. Biggs (1974). *Algebraic graph theory*. Cambridge Tracts in Mathematics, 67.
7. N.L. Biggs, E.K. Lloyd and R.J. Wilson 1976. *Graph Theory 1736-1936*. Oxford.
8. G.D. Birkhoff and D.C. Lewis (1946). Chromatic polynomials. *Trans. Amer. Math. Soc.*, **60**, 355–451.
9. D. Blanusa and S. Bilinski (1949). Proof of the indecomposability of a certain graph (Croatian). *Hrvatsko Prirodoslovno Drustvo Glasnik Mat-Fiz. Astr.*, Ser II, **4**, 78–80.
10. R.L. Brooks (1941). On colouring the nodes of a network. *Proc. Cambridge Phil. Soc.*, **37**, 194–197.
11. R.L. Brooks, C.A.B. Smith, A.H. Stone and W.T. Tutte (1940). The dissection of rectangles into squares. *Duke Math. J.*, **7**, 312–340.
12. R.L. Brooks, C.A.B. Smith, A.H. Stone and W.T. Tutte (1975). Leaky electricity and triangulated triangles. *Philips Res. Reports*, **30**, 205–219.
13. W.G. Brown (1963). Enumeration of non-separable planar maps. *Can. J. Math.*, **15**, 526–545.
14. W.G. Brown and W.T. Tutte (1964). On the enumeration of rooted non-separable planar maps. *Can. J. Math.*, **16**, 572–577.
15. T. Brylawski (1975). On the non-reconstructibility of combinatorial geometries. *J. Comb. Theory (B)*, **19**, 72–76.
16. Lewis Carroll The hunting of the snark.
17. P.A. Catlin (1979). Hajós' graph-coloring conjecture; variations and counterexamples. *J. Comb. Theory (B)*, **26**, 268–274.
18. H.S.M. Coxeter (1983). My graph. *Proc. London Math. Soc.(3)*, **46**, 117–136.
19. Blanche Descartes (1948). Network colourings. *Math. Gaz.*, **32**, 67–69.

20. G.A. Dirac (1952). Some theorems on abstract graphs. *Proc. London Math. Soc. (3)*, **2** 69–81.
21. A.J.W. Duijvestijn (1962). Electronic computation of squared rectangles. Thesis, Technological University, Eindhoven. Philips Res. Reports, 17, 523–612.
22. A.J.W. Duijvestijn (1978). Simple perfect squares of lowest order. *J. Comb. Theory B*, **25** 240–243.
23. A.J.W. Duijvestijn, P.J. Federico and P. Leeuw (1982). Compound perfect squares. *Amer. Math. Monthly*, **89** 15–32.
24. J. Edmonds (1965). Paths, trees and flowers. *Can. J. Math.*, **17**, 449–467.
25. P. Erdös and T. Gallai (1960). Graphs with prescribed valencies. *Mat. Lapok*, **11** 264–274. (Hungarian).
26. S. Foldes (1978). The rotor effect can alter the chromatic polynomial. *J. Comb. Theory (B)*, **25**, 237–239.
27. R. Frucht (1952). A one-regular graph of degree 3. *Can. J. Math.*, **4**, 240–247.
28. T. Gallai (1963). Neuer Beweiss eines Tutte'schen Satzes. *Magyar Tud. Akad. Mat. Kutato Int. Kozl.*, **9** 135–139.
29. P.M. Gibson (1972). The Pfaffians and 1-factors of graphs. *Trans. New York Acad. Sci. (2)*, **34** 52–57.
30. P.M. Gibson (1972). The Pfaffians and 1-factors of graphs II. In *Graph Theory and Applications*, pp. 89–98. Berlin.
31. E. Ya. Grinberg (1968). Plane homogeneous graphs of degree three without Hamiltonian circuits. (Russian: Latvian and English summaries). *Latvia Math. Yearbook* 4, pp. 51-58. Riga.
32. D.W. Hall (1974). Coloring seven-circuits. Lecture Notes in Math., Vol. 406, pp. 273–290. Springer, Berlin.
33. D.W. Hall and D.C. Lewis (1948). Coloring six-rings. *Trans. Amer. Math. Soc.*, **64**, 184–191.
34. D.W. Hall, J.W. Siry and B.R. Vanderslice (1965). The chromatic polynomial of the truncated icosahedron. *Proc. Amer. Math., Soc.*, **16**, 620–628.
35. P. Hall (1935). On representations of subsets. *J. London Math. Soc.*, **10** 26–30.
36. P.J. Heawood (1890). Map colour theorem. *Quart. J. Math.*, **24** 332–338.
37. P.J. Heawood (1949). Map colour theorem. *Proc. London Math. Soc. (2)*, **51**, 161–175.
38. Rufus Isaacs (1975). Infinite families of nontrivial trivalent graphs which are not Tait colourable. *Amer. Math. Monthly*, **82**, 221–239.
39. F. Jaeger (1976). On nowhere-zero flows in multigraphs. *Congressus Numerantium* **XV**, Utilitas Math., Winnipeg, Manitoba, 373–378.
40. P.W. Kasteleyn (1963). Dimer statistics and phase transitions. *J. Math. Phys.*, pp. 287–293.

41. N.D. Kazarinoff and P. Weitzenkamp (1973). On existence of compound perfect squares of small order. *J. Comb. Theory B*, **14**, 163–179.
42. W.L. Kocay (1981). An extension of Kelly's Lemma to spanning subgraphs. *Congressus Numerantium*, **31**, 109–120.
43. L. Lee (1974). Chromatically equivalent graphs. Ph.D. dissertation, George Washington University.
44. Liu Yanpei (1984). On the number of rooted c-nets. *J. Comb. Theory (B)*, **36**, 118–123.
45. F.G. Maunsell (1952). A note on Tutte's paper 'The factorization of linear graphs'. *J. London Math. Soc.*, **27**, 127–128.
46. W.F. McGee (1960). A minimal cubic graph of girth seven. *Can. Math. Bull.*, **3**, 149–152.
47. R.C. Mullin and P.J. Schellenberg (1968). The enumeration of c-nets via quadrangulations. *J. Comb. Theory*, **4**, 259–276.
48. J. Petersen (1891). Die Theorie der regulären Graphs. *Acta Math.*, **15** 193–220.
49. J. Petersen (1898). Sur le théorème de Tait. *L'Intermédiaire des mathématiciens*, **5**, 225–227.
50. L. Pósa (1963). On the circuits of finite graphs. *Publ. Math. Inst. Hung. Acad. Sci.*, **8** A3, 355–361.
51. M.S. Sainte-Laguë (1926). Les réseaux (ou graphes). *Mémorial des sciences mathematiques*, XVIII. Paris.
52. H. Seifert and W. Threlfall (1934). *Lehrbuch der Topologie*. Leipzig.
53. P.D. Seymour (1981). Nowhere-zero 6-flows. *J. Comb. Theory (B)*, **30**, 130–135.
54. C.A.B. Smith (1971). Map colourings and linear mappings. In *Combinatorial mathematics and its applications*, Proc. Conf. Oxford (1969) pp. 259-283. London.
55. R. Sprague (1939). *Mathematische Zeitschrift*, **45** 607.
56. R. Sprague (1940). *Mathematische Zeitschrift*, **46** 460–471.
57. P.K. Stockmeyer (1977). The falsity of the reconstruction conjecture for tournaments. *J. Graph Theory*, **1**, 19–25.
58. W.T. Tutte (1950). Squaring the square. *Can. J. Math.*, **2**, 197–209.
59. W.T. Tutte (1961). Squaring the square. In *2nd Scientific American Book of Mathematical Puzzles and Diversions*, by Martin Gardner. New York.
60. W.T. Tutte (1946). On Hamiltonian circuits. *J. London Math. Soc.*, **21**, 98–101.
61. W.T. Tutte (1956). A theorem on planar graphs. *Trans. Amer. Math. Soc.*, **182**, 99–116.
62. W.T. Tutte (1960). A non-Hamiltonian planar graph. *Acta Math. Acad. Sci. Hung.*, **11**, 371–375.
63. W.T. Tutte (1978). Bridges and Hamiltonian circuits in planar graphs. *Aequationes Math.*, **17**, 121–140.

64. W.T. Tutte (1947). The factorization of linear graphs. *J. London Math. Soc.*, **22**, 107–111.
65. W.T. Tutte (1952). The factors of graphs. *Can. J. Math*, **4**, 314–328.
66. W.T. Tutte (1954). A short proof of the factor theorem for finite graphs. *Can. J. Math.*, **6**, 347–352.
67. W.T. Tutte (1974). Spanning subgraphs with specified valencies. *Discrete Mathematics*, **9**, 97–108.
68. W.T. Tutte (1948). The dissection of equilateral triangles into equilateral triangles. *Proc. Cambridge Phil. Soc.*, **44** 463–482.
69. W.T. Tutte (1973). Duality and trinity. *Colloq. Math. Soc. J. Bolyai*, **10**, 1459–1472.
70. W.T. Tutte (1981). Dissections into equilateral triangles. In *The Mathematical Gardner*, ed. D.A. Klarner. Wadsworth International, Belmont, California, 127–139.
71. W.T. Tutte (1976). The rotor effect with generalized electrical flows. *Ars Combinatoria*, **1**, 3–31,
72. W.T. Tutte (1947). A ring in graph theory. *Proc. Cambridge Phil. Soc.*, **43** 26–40.
73. W.T. Tutte (1949). On the imbedding of linear graphs in surfaces. *Proc. London Math. Soc. (2)*, **51**, 474–483.
74. W.T. Tutte (1954). A contribution to the theory of chromatic polynomials. *Can. J. Math.*, **6**, 80–91.
75. W.T. Tutte (1980). 1-factors and polynomials. *Europ. J. Combinatorics*, **1**, 77–87.
76. W.T. Tutte (1947). A family of cubical graphs. *Proc. Cambridge Phil. Soc.*, **43**, 26–40.
77. W.T. Tutte (1959). On the symmetry of cubic graphs. *Can. J. Math.*, **11**, 527–552.
78. W.T. Tutte (1960). A non-Hamiltonian graph. *Can. Math. Bull.*, **3**, 1–5.
79. W.T. Tutte (1961). Symmetrical graphs and coloring problems. *Scripta Math.*, **25**, 305–316.
80. W.T. Tutte (1974). Codichromatic graphs. *J. Comb. Theory (B)*, **16**, 168–174.
81. W.T. Tutte (1948). Thesis, Cambridge.
82. W.T. Tutte (1965). Lectures on matroids. *J. Res. Nat. Bur. Standards, Sect. B*, **69B**, 1–47.
83. W.T. Tutte (1984). *Graph Theory*. Addison-Wesley, New York.
84. W.T. Tutte (1956). A class of Abelian groups. *Can. J. Math.*, **8**, 13–28.
85. W.T. Tutte (1958). A homotopy theorem for matroids, I. *Trans. Amer. Math. Soc.*, **88**, 144–160.
86. W.T. Tutte (1958). A homotopy theorem for matroids, II. *Trans. Amer. Math. Soc.*, **88**, 161–174.

87. W.T. Tutte (1959). Matroids and graphs. *Trans. Amer. Math. Soc.*, **90**, 621–624.
88. W.T. Tutte (1960). An algorithm for determining whether a given binary matroid is graphic. *Proc. Amer. Math. Soc.*, **11**, 905–917.
89. W.T. Tutte (1967). On dichromatic polynomials. *J. Comb. Theory*, **2**, 301–320.
90. W.T. Tutte (1979). All the King's horses. A guide to reconstruction. In *Graph Theory and related topics*, ed. ?? pp. 15-33. Academic Press, New York.
91. W.T. Tutte (1962). A census of planar triangulations. *Can. J. Math.*, **14**, 21–38.
92. W.T. Tutte (1962). A census of Hamiltonian polygons. *Can. J. Math.*, **14**, 402–417.
93. W.T. Tutte (1962). A census of slicings. *Can. J. Math.*, **14**, 708–722.
94. W.T. Tutte (1963). A census of planar maps. *Can. J. Math.*, **15**, 249–271.
95. W.T. Tutte (1968). On the enumeration of planar maps. *Bull. Amer. Math. Soc.*, **74**, 64–74.
96. W.T. Tutte (1969). On the enumeration of four-coloured maps. *SIAM J. Appl. Math.*, **17**, 454–460.
97. W.T. Tutte (1973). Chromatic sums for rooted planar triangulations: the cases $\lambda = 1$ and $\lambda = 2$. *Can. J. Math.*, **25**, 426–447.
98. W.T. Tutte (1973). Chromatic sums for rooted planar triangulations II: the case $\lambda = \tau + 1$. *Can. J. Math.*, **25**, 657–671.
99. W.T. Tutte (1973). Chromatic sums for rooted planar triangulations III: the case $\lambda = 3$. *Can. J. Math.*, **25**, 780–790.
100. W.T. Tutte (1973). Chromatic sums for rooted planar triangulations IV: the case $\lambda = \infty$. *Can. J. Math.*, **25**, 929–940.
101. W.T. Tutte (1974). Chromatic sums for for rooted planar triangulations V: special equations. *Can. J. Math.*, **26**, 893–907.
102. W.T. Tutte (1982). Chromatic solutions. *Can. J. Math.*, **34**, 741–758.
103. W.T. Tutte (1982). Chromatic solutions II. *Can. J. Math.*, **34**, 952–960.
104. W.T. Tutte (1970). On chromatic polynomials and the golden ratio. *J. Comb. Theory*, **9**, 289–296.
105. W.T. Tutte (1970). More about chromatic polynomials and the golden ratio. *Combinatorial structures and their applications*, pp. 439–453. (ed R. K. Guy et al.) Gordon and Breach, New York.
106. W.T. Tutte (1978). On a pair of functional equations of combinatorial interest. *Aequationes Mathematicae*, **17**, 121–140.
107. D. McCarthy and R.G. Stanton, eds. (1979). Selected Papers of W.T. Tutte. The Charles Babbage Research Centre, St. Pierre, Manitoba, Canada.
108. O. Veblen (1931). *Analysis Situs*. Colloquium Publications Amer. Math. Soc., Vol. V, Part II. New York.

109. H. Walther (1965). Ein kubischer planarer zyklisch fünffach zusammenhängender Graph der keinen Hamiltonkreis besitzt. *Wiss. Z. Techn. Hochsch. Ilmenau*, **11**, 163–166.
110. K. Wagner (1960). Bemerkung zu Hadwigers Vermutung. *Math. Ann.*, **141**, 433–451.
111. H. Walther and H-J. Voss (1974). *Über Kreise in Graphen*. Berlin.
112. D.J.A. Welsh (1976). *Matroid Theory*. Academic Press, London.
113. H. Whitney (1931). A theorem on graphs. *Ann. of Math. (2)*, **32** 378–390.
114. H. Whitney (1932). The coloring of graphs. *Ann. of Math. (2)*, **33**, 688–718.
115. H. Whitney (1932). A logical expansion in mathematics. *Bull. Amer. Math. Soc.*, **38**, 572–579.
116. H. Whitney (1933). 2-isomorphic graphs. *Amer. J. Math.*, **55**, 245–254.
117. H. Whitney (1932). Non-separable and planar graphs. *Trans. Amer. Math. Soc.*, **34**, 339–362.
118. H. Whitney (1935). On the abstract properties of linear dependence. *Amer. J. Math.*, **57**, 509–533.
119. T.H. Willcocks (1948). Problem 7795 and solution. *Fairy Chess Review*, **7** p. 106.
120. T.H. Willcocks (1951). A note on some perfect squared squares. *Can. J. Math.*, **3**, 304–308.

INDEX

1-Factor Theorem, 28–32
2-Connected graphs, 7, 111
2-Separable graphs, 7
3-Connected graphs, 7, 83, 91, 111, 144, 145
4-Connected graphs, 22
5-Flow Conjecture, 51, 52
c-Net of a squared rectangle, 4, 124
f-Factor Theorem, 31, 32
m-Colouring, 49, 50, 52
m-Flow, 51, 52
n-Clique, 82
s-Regular graphs, 76, 79, 80

Algebraic duality, 58, 85, 87
Algorithm for graphic matroids, 105
Alternating map, 36–39, 42, 43, 45
Alternating paths, 28, 30, 32, 33
Anacker, S., 111
Augusteijn, L., 43
Automorphism, 64, 75, 78, 80
Avoidance (bridges in a matroid), 103
Avoidance (bridges in graphs), 89, 92

Balanced directed network, 39
Bands, 121, 125
Bari, Ruth, 73, 133
Barnette's Conjecture, 44
Barycentric representation of planar graphs, 144
Beraha numbers, 136, 138, 139
Beraha polynomials, 146
Beraha, S., 135
Berman, G., 56, 133
Berman, K., 45
Bernhart, F., 73, 136
Bicubic map, 44, 45, 123
Biggs, N., 80, 109, 136, 145
Binary chain-groups, 95
Binary matroid, 100, 102
Birkhoff, G.D., 57, 129
Birkhoff-Lewis equations, 138, 145
Block, 92, 96, 104, 111
Bond, 86–88, 91, 92, 95, 104
Boundary, 49
Boundary of a chain, 47, 49
Bouwkamp, C.J., 143, 145
Bridges of a bond, 92, 104

Bridges of a circuit, 88, 92
Brooks' Theorem, 1, 57, 82, 83, 85
Brooks, Mrs., 140, 142
Brooks, R.L., 1, 140, 142
Brown, W.G., 124

Cage, 79
Canterbury Puzzles, 1
Cayley, A., 114
Cells, 95–98, 101
Cells of a matroid, 99
Chain-group, 95–104
Chains over a ring, 46
Characteristic polynomial, 109, 112, 113
Cheshire Cat, 94
Chess, 12
Chromatic eigenvalues, 127, 139
Chromatic number, 109, 111
Chromatic polynomial, 52, 54, 60, 72, 73, 107–109, 111, 125, 126, 129, 131, 133
Chromatic sums, 126, 127, 138, 139
Circuits of a matroid, 100
Coboundary, 52
Combinatorial Topology, 46, 47, 50, 94
Complexity at a vertex, 40
Complexity of alternating network, 40
Complexity of an electrical network, 8, 11, 57, 66
Component of a chain-group, 96
Components of a graph, 53
Compound perfect rectangle, 7
Constrained chromial, 129, 130
Contraction of a chain-group, 95
Contraction of a loop, 91
Contraction of an edge, 53, 91
Contraction of the last edge, 92
Coxeter, H.S.M., 12, 79, 80
Cremona–Richmond configuration, 79
Cross, 4, 43
Cubic graphs, 13, 18, 46, 51, 52, 58–60, 63, 74, 82, 94
Current, 3, 4, 7, 37, 38, 41, 43, 49, 143

Dahab, R., 146
Degenerate bridges, 23
Deletion of an edge, 53, 91
Dendroid, 97–99, 101, 102

Dendroid of a chain-group, 97
Dichromate, 54, 55, 57, 67, 70, 72, 73
Dichromatic polynomial, 53, 54, 110, 111
Differential equations, 128, 139
Dirac, G.A., 22
Directed edge, 36, 37, 48, 49
Directed network, 36
Directed tree, 40
Dodecahedral graph, 74
Dodecahedron, 133, 136
Dual graphs, 56, 92
Dual matroid, 100
Dual of a chain-group, 96
Duijvestijn, A.J.W., 143

Edge-reconstruction, 112
Electrical networks, 3, 4, 6–9, 11, 36, 38, 39, 42–44, 66
Elementary chain, 95–97, 103
Elementary coboundary, 86, 87
Elementary cycle, 47, 85, 96
Equations of generating functions, 118, 119, 126, 127
Euler Polyhedron Formula, 36
Eulerian planar maps, 123

Faces of a c-net, 4
Factors of a graph, 24, 60
Fano matroid, 101, 102
Flats of a matroid, 103
Flattening equations, 136
Flipping a rotor, 66, 70
Flow-polynomial, 53, 54, 60, 111
Foldes, S., 72
Foster, R.M., 53, 80
Four Colour Problem, 13, 81
Four Colour Theorem, 24, 25, 58, 85, 128
Four-Colour averages, 125
Free chromial, 130–132
Frucht, R., 80
Full elements of a squared rectangle, 8
Full flow in an electrical network, 41
Full sides of a squared rectangle, 8

Gallai, T., 30
Generating function, 116, 118–120, 125
Golden identity, 134–137
Golden number, 126, 134
Golden root, 135–137
Graphic matroid, 102
Grinberg's Theorem, 18, 19, 21, 87, 88
Grinberg's Theorem, dual form, 88
Grinberg, E., 18

Hadwiger's Conjecture, 85
Hajós' Conjecture, 85
Hall's Theorem, 30
Hall, D.W., 133–136
Hall, P., 30
Hamilton, Sir W.R., 12
Hamiltonian bond, 87, 88
Hamiltonian circuit, 1, 13–19, 21, 22, 24, 26, 33, 44, 47, 74, 87, 109, 111, 112, 120
Hamiltonian cycle, 47
Heawood graph, 77, 78
Heawood, P.J., 44, 46
Honsberger, R., 144
Hyperprime graph, 29

Impedances, 73
Integral chain-group, 95
Isomers, 141
Isthmus, 25, 51, 54, 55, 67, 69, 70, 91, 116

Jacobi's Theorem, 26, 27, 73, 113
Jaeger, F., 51
Jordan's Theorem, 88

Kasteleyn, P.J., 27, 63
Kelly's Lemma, 109–113
Kirchhoff matrix, 7, 26, 27, 99, 112
Kirchhoff matrix generalized, 39
Kirchhoff's Laws, 7, 140
Kirchhoff's Laws generalized, 37
Kirkman, T.P., 13, 145
Knight's Tour, 12, 13
Kocay, W.L., 110
Kuratowski's Theorem, 76, 89, 103

Leaky electricity, 34, 37, 40
Lee, L., 73
Lewis, D.C., 133
Lines of a matroid, 103
Link-map, 125, 126
Loop, 21, 28, 30, 49, 52, 54, 55, 59, 60, 67, 69, 70, 85, 91, 92, 107, 116, 131
Loose edge, 59, 60
Lusin's Conjecture, 2, 11

Map, 13, 24, 75, 82
Matrix of chromatic joins, 146
Matrix–Tree Theorem, 11, 40, 43, 44, 99
Matrix–Tree Theorem for directed graphs, 40

INDEX

Matroid, 99–102, 145
Matroid of a chain-group, 102
Maunsell, F.G., 28
McGee, W.F., 79
Minor of a chain-group, 102
Minor of a graph, 85
Minors of a matroid, 102
Moroń, Z., 2
Mullin, R.C., 117, 124

Near-triangulations, 22, 116, 118, 125–127, 138
Non-Hamiltonian cubic planar maps, 21
Non-Hamiltonian graphs, 79
Non-Hamiltonian maps, 18, 19
Nowhere-zero coboundary, 52
Nowhere-zero cycle, 50–52

Order of a squared rectangle, 3
Order of a triangulated parallelogram, 35
Orthogonal chains, 58
Outgrowths, 92
Overlapping bridges in graphs, 89, 93
Overlapping bridges in matroids, 103

Parametric equations, 119, 120, 124, 139
Partition of a bond, 92
Partition of the vertex-set, 129
Perfect rectangles, 2–4, 6, 8, 10, 35, 140, 141
Perfect squares, 10, 11, 26, 64, 67, 142, 143
Perfect triangulations, 35
Peripheral circuits, 89, 91, 104, 105
Peripheral circuits in a matroid, 104
Petersen graph, 25, 26, 51, 75–77, 79, 85
Petersen's Theorem, 1, 25, 26, 29, 30
Petersen, J., 24, 25
Pfaffian, 26, 27, 63
Planar graphs, 10, 27, 58, 63, 81, 82, 87, 145
Planar partition, 136
Planar triangulations, 22, 114, 133–135
Plane graphs, 81, 144
Points of a matroid, 103
Polar edges, 4
Polyhedra, 7, 13, 81, 82, 124
Prime graphs, 27
Primitive chain-groups, 97
Prism, 14, 17, 18
Pósa, L., 22

Radiants of a face, 17
Rank of a chain-group, 97
Reconstructible graph, 106
Reconstructible graph-property, 106
Reconstruction Conjecture, 106, 112, 113
Reduced sides and elements, 8
Reduction, 8, 26, 41, 42, 140
Reduction of a chain-group, 95, 97
Reduction of a squared rectangle, 8, 26, 142
Regular chain-group, 97, 99–101
Regular elements of a ring, 97
Regular graphs, 24
Regular matroid, 102
Representative matrix of a chain-group, 99
Residual arcs of a bridge, 92
Residual graph of a circuit, 22
Rooted 3-connected planar maps, 124
Rooted non-separable planar maps, 124
Rooted planar maps, 124
Rooted triangulations, 116, 125, 128
Rooted triangulations, with colouring, 125
Rotors, 64, 65, 70, 72, 140, 141

Sachs, H., 23
Sainte-Laguë, M.A., 26
Sands, D., 136
Schellenberg, P.J., 124
Semiperimeter, 8
Separable graphs, 6, 56
Separating digon, 114
Separating triangle, 22, 114
Serpens, 74–76, 79
Seymour, P.D., 51
Shearing a squared rectangle, 34
Silver root, 135, 137
Simple perfect rectangle, 7
Simple perfect square, 11, 143
Sims, C., 80
Skinner, J.D., 143
Slicing of the band, 121
Smith's Theorem, 1, 18, 47, 48, 57, 94
Smith, C.A.B., 1, 13, 44
Snark, 51, 52, 85
Spanning tree, 11, 40, 43, 44, 98
Spectrum of a graph, 112
Sprague, R.P., 10, 142
Squared rectangle, 3, 4, 7, 8, 14, 34, 36, 38, 39, 44, 113, 143
Standard representative matrix of a chain-group, 99, 101
Stator, 65–67, 69, 70, 74

Steinitz' Theorem, 7, 13
Stirling's formula, 120
Stone, A.H., 1
Support of a chain, 85

Tait colouring, 24, 25, 47–51, 79, 85, 94
Tait cycle, 24–26, 47, 48, 51, 94, 121
Tait's Conjecture, 13–15, 17–19
Tait, P.G., 13, 24, 82
Theoretical perfect squares, 10
Theory of numbers, 142
Topological invariance, 60
Totally unimodular matrix, 99, 101
Transpedance, 73
Tree-number, 67, 73, 74
Trees, 11, 24, 27, 40, 87, 109, 113, 114
Triangulated parallelograms, 34–38, 40, 42, 43, 45
Triangulated triangles, 34, 36, 42–45
Trine alternating graphs, 44
Trinity College, Cambridge, 13
Trivial imperfection in a c-net, 143
Truncated icosahedron, 133–136
Tutte polynomial, 54, 57
Twisting an edge in a cubic graph, 58

Ulam's Conjecture, 106
Unsymmetrical electricity, 40, 48

V-Functions, 60
Veblen, O., 46
Vertex-deleted subgraphs, 106, 107, 109, 112
Vertex-elimination Theorem, 136
Vertex-graph, 54, 92, 106, 107
Vertices of a bridge, 89
Vertices of attachment of a bridge, 23, 89
Vertices of attachment of a rotor, 9
Voltage in an electrical network, 4

Walther, H., 18, 22
Wheel (graph), 145
Wheel (matroid), 145
Whirl (matroid), 145
Whitney's Theorem on 3-connected planar graphs, 91
Whitney's Theorem on near-triangulations 22
Whitney's Theorem on planar triangulations, 22
Whitney, H., 7, 22, 54, 99
Willcocks, T.H., 10, 143

Younger, D.H., 21, 146

Zaks, J., 21
Zeros of chromatic polynomials, 133

DATE DUE

NOV 6 '08			
DEC 0 4 2007			
		SUBJECT TO	
		RECALL	

Demco, Inc. 38-293